Lost Person Behavior

Instructor Notes 2014

Robert J. Koester

dbS Productions
Charlottesville, Virginia

Copyright © 2014 dbS Productions LLC
All Rights Reserved.

No part of this book may be reproduced in any form or by any electronic means including storage and retrieval systems without permission in writing from the publisher except by a reviewer who may quote brief passages in a review. For information contact dbS Productions.

Published by dbS Productions LLC
P.O. Box 94
Charlottesville, Virginia USA 22902
+1.434.293.5502
www.dbs-SAR.com

Printed in the United States

10 9 8 7 6 5 4 3 2 1

Publisher's Cataloging-in-Publication Data

Koester, Robert J. (Robert James), 1962-
Lost person behavior : Instructor Notes 2014 / Robert J. Koester
ISBN 978-1-879471-58-0 (pbk : alk. Paper)

Table of Contents

Course Goal 1
Course Objectives 2
Target Audience 3
Course Limitations 4
Course Structure 5
Duration 6
Instructor Qualifications 7
Overview of Course Units 8
Sample Agenda 9
Course References 10
Instructor Methodology 11
Preface 12

Unit 1 – Introduction 13
Unit 2 – International Search & Rescue Incident Database (ISRID) 31
Unit 3 – Lost Person Strategies 47
Unit 4 – Myths and Legends 67
Unit 5 – Using the Book 85
Unit 6 – Reflex Tasking 101
Unit 7 – Subject Categories 117

Instructor's Ordering Form 211

Plan of Instruction

COURSE GOAL

The goal of the course is to provide participants with the tools and knowledge necessary to successfully look in the right place, understand lost person behavior, apply statistical tables, identify needed special investigation, and deploy resources into the field. Participants should also be able to brief other searchers on key components of lost person behavior.

COURSE OBJECTIVES

Upon successfully completing this course, the participants will be able to:

- Describe key developments in the field of lost person behavior. (Unit 1)

- Describe how the ISRID database is organized to best predict behavior. (Unit 2)

- Demonstrate the ability to determine correct subject category given an actual incident scenario. (Unit 2)

- Judge different methods to combine different subject categories. (Unit 2)

- Explain different causes of becoming lost and different strategies lost persons attempt. (Unit 3)

- Demonstrate ability to identify decision points given a map and a lost person scenario. (Unit 3).

- Describe key components and common misconceptions of lost person behavior. (Unit 4)

- Use the correct statistical summary data from the ISRID database to best model a subject's possible location and survivability. (Unit 5)

- Demonstrate the use of reflex tasking to generate initial tasks given a map and a scenario. (Unit 6)

- Describe key points for each subject category profile. (Unit 7)

- Demonstrate the ability to deploy resources appropriate for each subject category given a map and scenario information. (Unit 7).

TARGET AUDIENCE

The target audience for this course is as follows:

Primary Audience	Secondary Audience
Search and Rescue (SAR) Management	Public Safety Officials
Search and Rescue (SAR) Field personnel	Public Safety Responders
Law Enforcement personnel	Parks and Recreations participants
First Responders	Education and Academic participants

The lost person behavior course has no prerequisites. Those with a search and rescue background will get the maximum benefit from the course.

COURSE LIMITATIONS:

This course is not intended to replace or be in lieu of any search management course. While it provides effective tools to aid in determining where to look and provides initial tactics, several critical search management concepts and theory are not addressed in this training. The actual methods to task teams using the correct tactics are not covered. In addition, critical concepts not addressed include but are not limited to; pre-planning, SAR resources, resource management, management systems, legal issues, risk assessment, search urgency, strategy, formal search theory, consensus methods, clue management, incident action plans, documentation, search suspension, etc.

COURSE STRUCTURE/STRATEGY

All participants will be prepared for an environment of interactive lectures, class participation, giving mini-presentations, and working independently or in groups to complete activities. Participants will be encouraged to apply their existing search and rescue skills and knowledge as well as those newly acquired in challenging and dynamic scenarios. Each unit will have a lecture portion, and discussion opportunities, accompanied by group activities to highlight, expand, and practice the teaching points of the course.

An important part of this course will be the organization of students into small groups of two to four people. This will ensure the maximum participation in the group discussions and activities. Course instructors will help facilitate group activities and discussions. These activities include opportunities for mentoring, providing immediate feedback and ensuring that group activities are performed as specified in the course objectives.

Students will be required to demonstrate their acquisition of the skills and knowledge through activities revolving around a given search scenario. Throughout the course, students will be given various search scenarios.

DURATION

This course is designed for either a 1-day format requiring 7 student hours or a two-day format requiring 14 student hours. This time includes lecture, group activities, breaks, and the end-of-course exam. It does not include the lunch break. The course may be delivered all at once, or broken up into units and delivered over a longer period of time.

If the map problems are being used as a stand alone training component the time period will be variable and no course certification will be offered.

Course Delivery

The course will be delivered as a one or two day workshop. Some instructors may also elect to teach the course in hour long blocks of time over an extended time.

The course is designed to meet the requirements of the International Association for Continuing Education & Training (IACET) standard (ANSI/IACET 1-2007 Standard). The one-day class will award 0.7 CEU and the two-day class will award 1.4 CEU provided the course is not shortened and meets all other requirements. Awarding CEU is dependent upon several factors such as qualified instructors, following the plan of instruction, proper documentation of the student's attendance, and successful completion of the learning objectives.

INSTRUCTOR/FACILITATOR QUALIFICATIONS

For field offerings the course will be managed by a Lead Instructor who is responsible for scheduling and managing the overall course delivery.

This course is designed for delivery by dbS Productions Certified Instructors who have search and rescue experience, search management experience, proven instructional experience, have successfully completed a Lost Person Behavior (LPB) workshop, and have successfully completed a Lost Person Behavior Train-The-Trainer workshop. Experienced search and rescue trainers may also deliver specific sections of instruction in their sphere of competence should the need arise. **Only**

certified instructors may deliver the Lost Person Behavior course. Any instructor may use the map exercises as a standalone.

It is recommended that a Lead Instructor be selected from the pool of instructors.

The Lead Instructor should be able to:

- Assist the dbS Productions course manager with resident or offsite deliveries.
- Provide the class prompt feedback on subject matter issue resolutions
- Serve as a map problem leader during group activities
- Facilitate discussion of subject issues arising among the instructor group
- Facilitate discussion of the exams and resolve any exam issues relating to the accuracy of the content
- Establish a contact with dbS Productions to discuss any factual or content issues.

Instructors will ensure that they:

- Are familiar with all course materials
- Have a copy of the course agenda
- Update their unit examples to remain timely
- Are current with their instructional skills

OVERVIEW OF COURSE UNITS AND EVALUATION

📖 **Unit 1: Introduction** This section provides information about the course location, the course's organizational structure, the instructors and participants. An overview of previous lost person behavior research is presented. A pretest will be given. The pretest explores both the participant's factual knowledge and ability to work a map problem.

📖 **Unit 2: International SAR Incident Database (ISRID)** provides participants an overview of the basic organization of the ISRID database. This allows the participant to use the correct statistical tables in order to best predict the behavior of a missing subject. The criterion for placing a missing subject into a subject category is shown. Participants are given two different actual scenarios to demonstrate the ability to correctly determine the correct subject category. Overall findings from the ISRID database regarding survivability and survivability factors are also shown.

📖 **Unit 3: Lost Person Strategies** provides participants information about the definition of lost, different scenarios that may cause a subject to become missing, and different strategies lost persons use when lost. Several scenarios are then presented which allow the participants to discuss the correct definition of lost and determine the appropriate scenario. The concept of decision points is introduced. Decision points are then illustrated with maps and photographs. Participants are then given several map problems that illustrate different strategies used by lost subjects. In the final section, participants working in groups must demonstrate the ability to identify decision points given a map and a lost person scenario.

📖 **Unit 4: Myths and Legends:** The instructor will discuss several common misconceptions about lost person behavior and statistical concepts. Topics include turning behaviors, uphill versus downhill, practical implications of statistical concepts, affects of climate on behavior, and computer modeling. At the end of this lesson, participants will be able to score the pre-test.

📖 **Unit 5: ISRID Tables explained** provides an overview of the statistical tables and models used in the ISRID database. The instructor will discuss with participants how the book Lost Person Behavior is organized. Then each of the predictive models (ring model, dispersion angle, elevation model, track offset model, mobility model, and feature model) will be demonstrated. The instructor will lead a discussion on how the different models may be integrated while planning. The models will be put into context with the current method of determining probability of containment (or probability of area) which is the Mattson methods. The instructor will also demonstrate the survivability statistics. Participants given a scenario will use the textbook to determine the subject's statistical location and potential survivability.

📖 **Unit 6: Reflex Tasking** The instructor will present the key components of reflex tasking using the bike model to organize basic types of tactical operations. Participants working in groups will then demonstrate the use of reflex tasking to generate initial tasks given a map and scenario information.

📖 **Unit 7: Subject Categories:** The instructor will present key concepts and highlights from each selected subject category. The instructor will tailor the presentation to meet the needs of participants based upon any regional differences. Instructor will foster class discussion by encouraging presentation of relevant cases from the local area. Instructor may also substitute map problems presented in the course with local map problems that demonstrate the same key concepts of lost person behavior. Participants should be able to use the textbook to determine the definition of each category, what types of activities or conditions included in each category, key profile points, the selection of relevant statistics, appropriate reflex tasks, and detailed investigation questions needed for each subject category. Several map problems are presented so that the participants can demonstrate the ability to deploy resources appropriate for each subject category given a map and scenario information. Potential subject categories to cover include; Abduction, Aircraft, Angler, ATV, Autistic, Camper, Caver, Child, Climber, Dementia, Despondent, Gatherer, Hiker, Horseback Rider, Hunter, Mental Illness, Mental Retardation, Mountain Biker, Other, Runner, Skier, Snowboarder, Snow-Mobiler, Snowshoer, Substance Abuse, Vehicle, Water, and Worker. The unit will be summarized with a brief discussion of the importance of collecting data.

📖 **Unit 8: Student Evaluation** The instructor will give students a written evaluation tool. The instructor may also elect to give the test instrument orally.

SAMPLE AGENDAS

A sample agenda is provided to assist the Course Manager or Lead Instructor to prepare for delivery of the two-day version of the course.

Two-Day Agenda

Day 1

1. Welcome and Opening	08:00 – 09:00
2. International SAR Incident Database (ISRID)	09:00 – 10:00
3. Lost Person Strategies	10:00 – 12:00
Lunch	12:00 – 13:00
4. Myths and Legends	13:00 – 14:15
5. ISRID Tables explained	14:15 – 15:15
6. Reflex Tasking	15:15 – 17:00

Day 2

7. Subject Categories — 08:00 – 12:00
Abduction, Aircraft, Angler, ATV, Autistic, Camper, Caver, Child, Climber, Dementia, Despondent, Gatherer, Hiker, Horseback Rider, Hunter, Mental Illness, Mental Retardation, Mountain Biker, Other, Runner, Skier, Snowboarder, Snow-Mobiler, Snowshoer, Substance Abuse, Vehicle, Water, Worker

Lunch	12:00 – 13:00
7. Subject Categories continued	13:00 – 16:30
8. Student Evaluation	16:30 – 17:00

ONE DAY AGENDA

Day 1

1. Welcome and Opening	08:30 – 9:15
2. International SAR Incident Database (ISRID)	9:15 – 11:00
3. Lost Person Strategies	10:00 – 11:00
4. Myths and Legends	11:00 – 12:00
Lunch	12:00 – 13:00
5. ISRID Tables explained	13:00 – 13:45
6. Reflex Tasking	13:45 – 14:45
7. Subject Categories	14:45 – 17:00

COURSE REFERENCES

The materials listed below are used in this course.

- *Lost Person Behavior* by Robert J. Koester*
- *Lost Person Behavior Student Workbook*
- *Lost Person Behavior: Instructor Activity Guide*
- *Lost Person Behavior: Instructor Notes*
- *Urban Search* by Chris Young and John Wehbring
- *Search Wheel*
- *Lost Person Behavior APP*
- Course Handouts
    - Map problems (if not included in student workbook)
    - Reflex Tasking Worksheet
    - Subject Category Wizard
    - Course administration (Agenda, class roster)
    - Student Evaluation Forms (one set per group instructor)

Note that for each offering, instructors must review current dbS Production publications to ensure that students are receiving the most recent version of the referenced documents.

*Each student is required to already have, purchase, or be given a copy of *Lost Person Behavior* as a key element of the instructor – dbS Productions presentation license agreement. If using this book as a stand-alone for map problems only the presentation license agreement does not require the purchase of *Lost Person Behavior*. However, it is highly recommended.

INSTRUCTOR METHODOLOGY

About the Instructor's Outlines

The material, as presented in the instructional outlines and slides, is not designed to be read verbatim. While this approach will work, the well prepared instructor must be able to tailor the presentation to the unique requirements of the particular audience. In addition, each instructor should have there own unique presentation style. The slides are simply a synopsis of main points. The instructor should have sufficient knowledge to give additional insight. While this instructor's notes does supply supplemental knowledge, instructors are strongly encouraged to be thoroughly familiar with the lesson plan, student manual, suggested web sites, required textbook, and suggested reading. Space is provided for the instructor to make special notes in the lesson plan.

Departures from the teaching material

The instructor's manual was created to serve as a resource manual. All material can be supported by referenced scientific papers. However, the author acknowledges that instructors may wish to make modifications to meet the needs of their audience. The decision to drop specific slides or sections does not require any acknowledgement. For certified instructor's additional case history slides may and should be added that are specific to the audience. However, additions to course content are not allowed unless approved by dbS Productions.

For instructor's using the map problems as stand-alone instruction no approvals are required from dbS Productions since the training is not offered as an actual course.

Presentation Methods:

Several different instructional strategies may be used to present the map problems. The most appropriate method will depend upon time, experience of the participants, and the homogeny of the students. The map problems may be presented as a class exercise, group exercise, or individual exercise. Ideally, different techniques should be used during the class.

Class Exercise

The problems may be presented as a class exercise. The instructor should inform the class they will be responsible for collecting investigative information and deploying resources. Either the PowerPoint should be shown or the initial map and assignment passed out to the class. Ideally, both should be used together. The instructor should read the initial description. The class should then be asked to either ask investigative or deployment questions. Information is collected by asking additional questions with the answers provided by the instructor. Resources are deployed by a student stating exactly where to send the resource. The students should respond to the effect; "search the residence" the instructor would respond "nothing found." This technique moves the class along the fastest. It also may allow dominant and or more experienced participants to provide most of the feedback. Several alternative instructional styles may also be selected.

Group Exercise
The instructor may also elect to break the class into groups. Each group should determine investigative questions it wishes to ask. After all groups are allowed the opportunity to ask questions each group should write on their group maps where they wish to deploy resources. Another method is to conduct the investigative phase as a class then the tactical aspect in groups.

Individual Exercise
Finally at least one of the reflex tasking map problems should be done individually (as they often are planned). However, the additional investigative information is best done as a class unless several assisting instructors are present. The instructor may also then pair students back into two person groups and the entire group to determine how to combine their planning. This often helps to illustrate why planning alone may have to happen, but planning together almost always enhances the overall SAR plan. Mixing different instructional techniques is typically the most effective.

Additional Comment about Map Problems: All the selected searches occurred from 1986-2009. Most of the incidents tend to be older. The passage of time was selected to insure a frank discussion is possible, since the statue of limitations has passed for any lawsuits. However, several changes in the SAR community have occurred. Development of higher standards for tracking dogs and mantrackers is a relatively new development. For this reason on many of the searches mantrackers and tracking dogs were not available at the time. Students at the completion of this course should request the deployment of these resources. However, in the interest of fairness, the students should be told "the resource is not available" versus making up bogus clues or trails.

Preface

Lost Person Behavior is a critical element of search theory. It provides information that helps the search planner, team leader, and team member's best determine where to look. If resources are not deployed to the correct location, the subject will not be found. If team leaders fail to deploy the team within the search area in an appropriate manner, the chances of making the find are reduced. Ultimately, it is the team member who must typically place their eyes on the subject, recognize the subject, and make the all important detection. Lost Person Behavior is one of the most important tools to help locate the subject.

A solid understanding of lost person behavior assists in the placement of resources into the correct location sooner. While this does not guarantee a find, it does increase the subject's odds of being located sooner. Faster finds results in fewer resources, less cost, and most importantly of all, a better chance of survival for the subject.

Field experience has shown that an in-depth understanding of lost person behavior can significantly reduce the time it takes to locate a missing subject. However, the topic is only offered in search management courses, and in a highly abbreviated format. This course provides guidance and formal training that teaches industry specialists to utilize this important tool.

This course is focused on lost person behavior primarily in a ground search and rescue environment. However, some sections do address missing aircraft and search subjects who enter the water from the ground. It does not address traditional maritime search and rescue. This course is intended to benefit new or experienced search and rescue team members, team leaders, and search planners. This course is not intended to provide in-depth instruction on search management in general.

After completing this course, participants will possess the basic knowledge necessary to apply the concepts of lost person behavior both in the field and in base work. In addition, search management personnel will have a better understanding of tools available to deploy the initial tasks.

The material in this book may also serve as standalone training or continuing education. Each of the various map problems or exercise could make the perfect practical problem for a team's regular training schedule.

Unit 1 Welcome and Introduction

Slide 1

Instructor should modify welcome screen to show the affiliations they hold. The instructor(s) may also modify screen to include their name(s) and perhaps the sponsor of the course. The LPB, ISRID, and dbS Logos may be used per the instructor dbS agreement.

Slide 2

This version of the start screen is for use by Robert Koester and is shown as an example of showing an instructor's affiliations.

Slide 3

Can also mention everyone is responsible for determining where to look. From management that places teams to the field resources that determine areas that need extra attention within in the search area to the actual searcher who ultimately decides where to actually look.

Unit 1 Welcome and Introduction

Slide 4

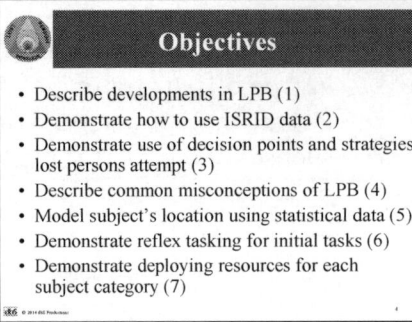

The objectives for the one and two day course are the same. The difference is the amount of time and depth spent on each unit.

Slide 5

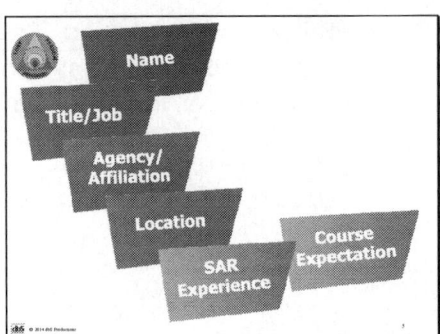

Each participant should provide the following. In large classrooms this may be skipped.

Slide 6

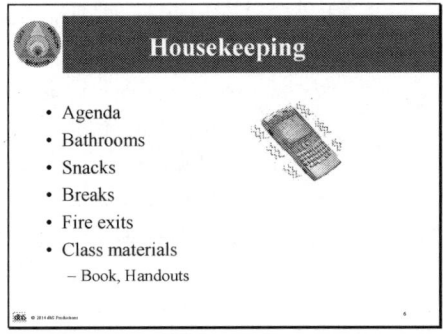

Review classroom management, inform class policy regarding phone use

Unit 1 Welcome and Introduction

Slide 7

Agenda may be modified to meet specific needs.

Slide 8

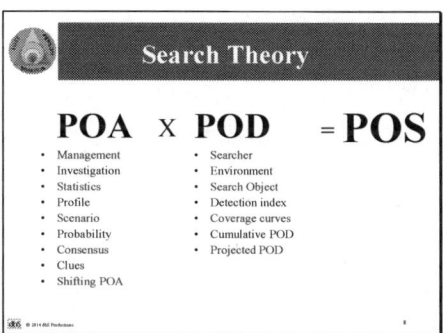

Slide should also be used to point out the elements that the course does not teach. This course is a bridge between field and more formal management courses. Be careful with SAR acronymns. This is a chance to point out the abbreviations section of the book found on page 359.

Slide 9

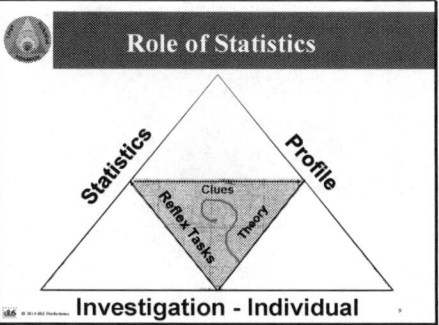

Triangle of information needed to first define search area, then inner triangle needed to actually find the subject. Slide is animated.

Unit 1 Welcome and Introduction

Slide 10  Nadia case used to show basic role of statistics and lost person behavior presented as a case.

Slide 11 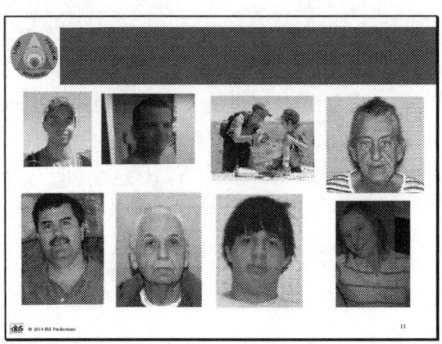 Potential SAR cases in background. Top left to right, then bottom left to right. Hiker, white water rafter, 8 year old boy, dementia, hiker, despondent with mental illness, Autism, and despondent

Slide 12 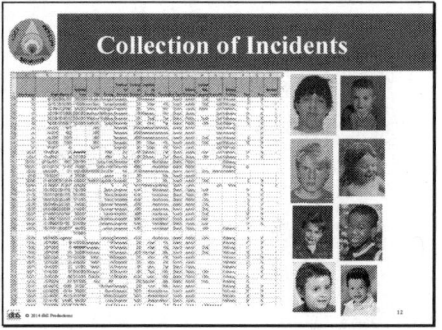 Lost Person Behavior approach collects many similar cases. Actual data looks dry, but recall each record represents actual incident.

Lost Person Behavior — September 2014

Unit 1 Welcome and Introduction

Slide 13

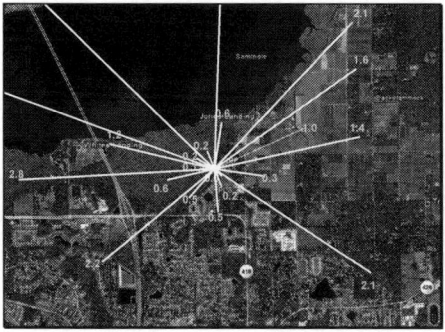

Plotting distances from the database like this would be unusual. However, it should be possible to begin to see a pattern emerge.

Slide 14

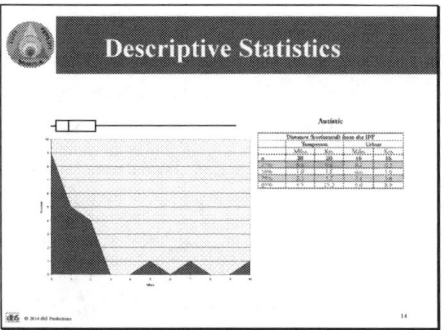

Different formats for showing statistics. The quartile approach is what is used in the Lost Person Behavior book.

Slide 15

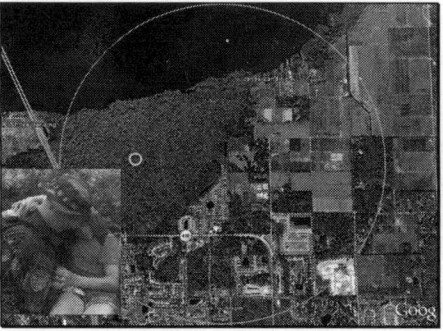

Most common procedure in SAR is to draw the quartiles rings on the map. In this example only the 25%, and 50% fit on the map. Subject was found just outside the 25% ring.

Lost Person Behavior

Unit 1 Welcome and Introduction

Slide 16

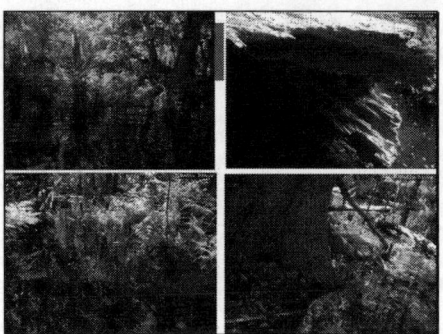

Pictures taken by Nadia which help to demonstrate attraction to lights, reflection, and water

Slide 17

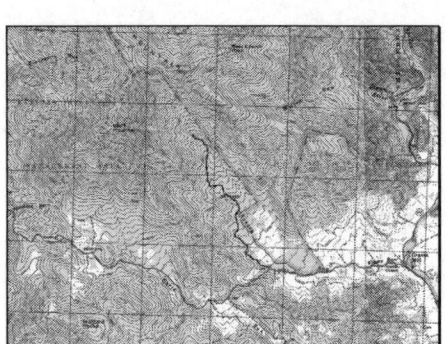

This map problem is presented to illustrate how it is still possible to carryout meaningful search planning even when no subject category or subject category statistics are available. At the time of this search the gatherer category had not been developed. See map problem handout "Unit 1 – Ginseng Hunter" for additional information on the incident. In brief, missing ginseng hunter last seen at home, knows area well, hunting ginseng entire life. Has been known to poach into national park. Has heart condition, complained of chest pain at church earlier in the morning. Last seen by wife, said he was going ginseng hunting. Expected to be back for dinner.

Unit 1 Welcome and Introduction

Slide 18

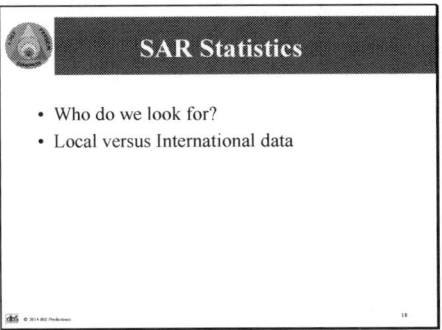

Databases give us an idea the various types of subjects and their behaviors that we look for. Ideally, lots of local data exists, this will best reflect what someone lost in your specific area will do. However, in many cases no data or not enough data exists, or not enough data for the specific subject category. In these cases International data is the best source.

Slide 19

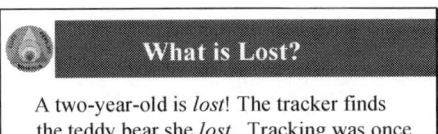

Different definitions of the word lost. When we talk about lost we typically use the first definition. Keep in mind that lost can only be defined by the subject. From the searcher perspective we only know the person is missing.

Slide 20

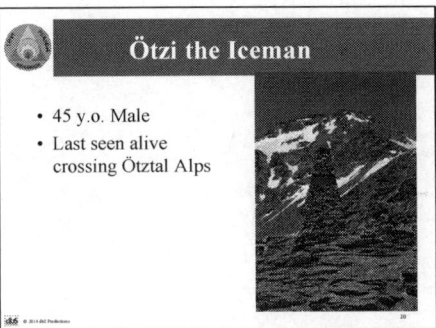

5300 years ago

Lost Person Behavior   September 2014

Unit 1 Welcome and Introduction

Slide 21

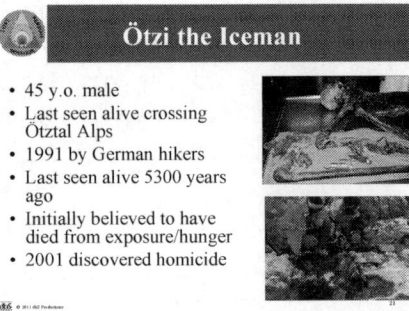

Need for law enforcement involved in Search and rescue, All searches a potential crime, Need to preserve find site as a crime scene, through secondary survey in the field needed.

Slide 22

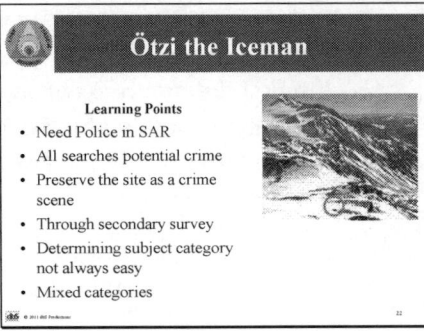

Need for law enforcement involved in Search and rescue, All searches a potential crime, Need to preserve find site as a crime scene, through secondary survey in the field needed.

Slide 23

Lost Person Behavior — September 2014

Unit 1 Welcome and Introduction

Slide 24

Slide 25

Examples used to show some of the differences with groups. Groups are important for survival. However, currently not shown in book Lost Person Behavior. M = solo males, F = solo females, MM = Group of all males, FF = group of all females, MF mixed gender group, AC = At least one adult and at least one child, gender neutral

Slide 26

Lost Person Behavior   21   September 2014

Slide 27

Slide 28

Hospice on the Plain of Jupiter better known as St. Bernard's Pass

Slide 29

Unit 1 Welcome and Introduction

Slide 30

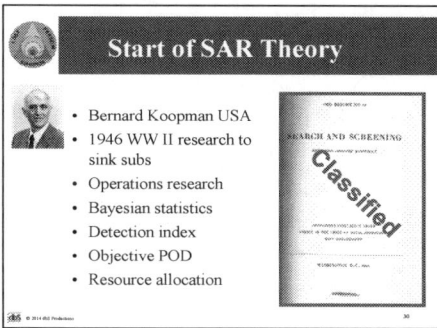

Start of Modern history of SAR. Almost all components of modern search theory were developed to locate, detect, and then sink subs.

Slide 31

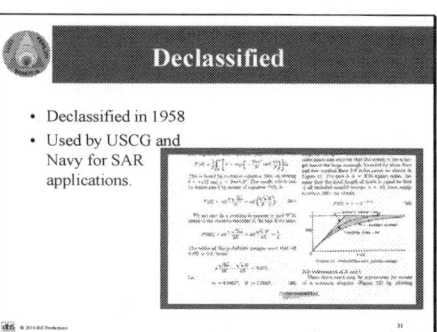

Search Theory has been used extensively by the USCG and US Navy ever since WWII. While mathematically intense to most people, it is well proven and applies on land as well. However, making it accessible to search planners is a challenge. The course is based upon modern search theory but taught in such a way to rely on rules of thumbs that work well during the initial stages of a search.

Slide 32

Cast of SAR behavior researchers. May be used in place of going through each individual which is presented in the add on slides.

Lost Person Behavior — September 2014

Unit 1 Welcome and Introduction

Slide 33

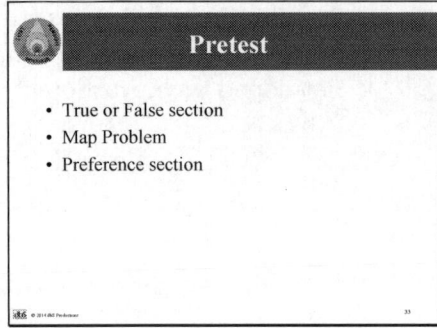

Slide 34

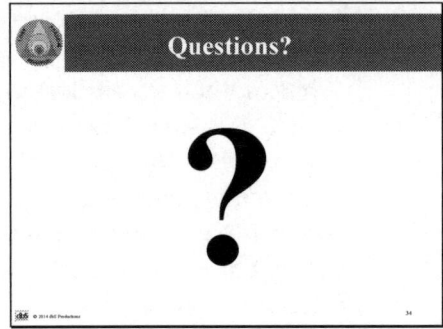

Slide 35

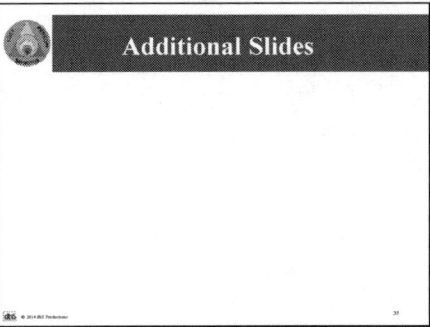

Lost Person Behavior — September 2014

Unit 1 Welcome and Introduction

Slide 36

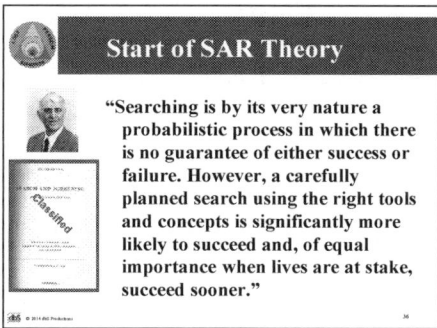

Optional Slide showing quote from Koopman about early SAR theory.

Slide 37

Animated slides used to illustrate how POA, Pden, and POD was used by Koopman. The four German sub bases are shown. The usual tracks away from the bases are then shown. The pden is the highest right at the base, but the red shows where they had german fighter aircraft cover. So where to search? The blue shows the search grid. It was also arranged to be at a 45 degree angle to the path since this had the highest POD.

Slide 38

Lost Person Behavior   25   September 2014

Unit 1 Welcome and Introduction

Slide 39

Slide 40

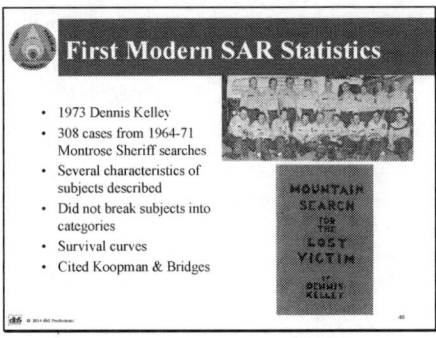

Slide 41

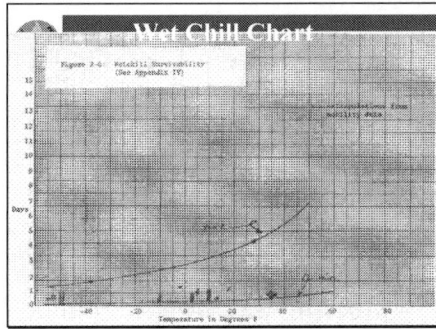

Lost Person Behavior   September 2014

Unit 1 Welcome and Introduction

Slide 42

Slide 43

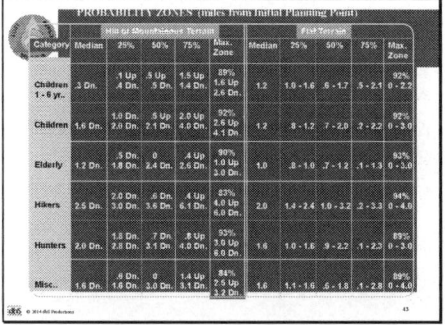

Slide 44

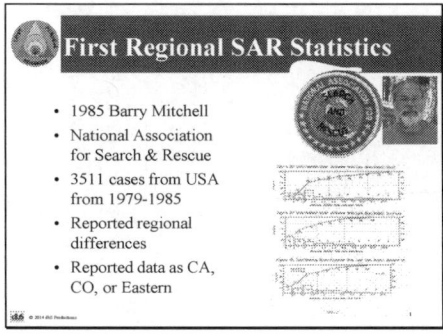

Lost Person Behavior   September 2014

Slide 45

Slide 46

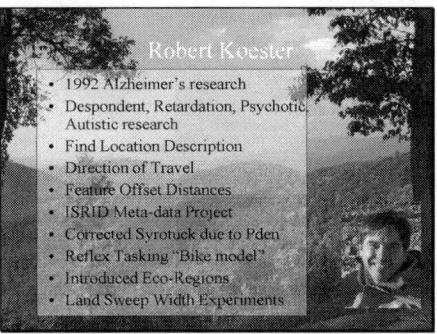

Slide 47

Unit 1 Welcome and Introduction

Slide 48

Slide 49

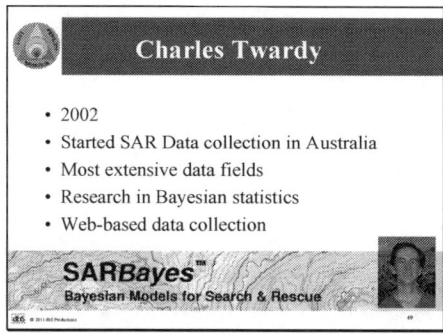

Slide 50

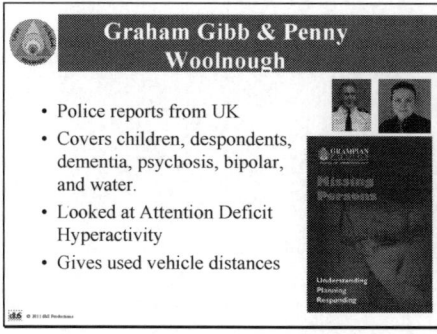

Unit 1 Welcome and Introduction

Slide 51

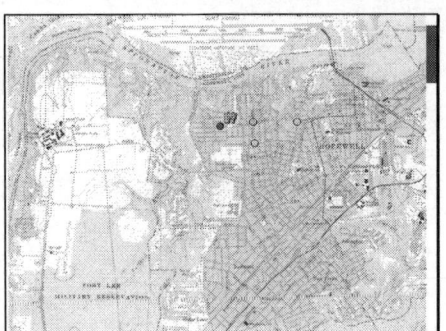

68 Y.O. male suffers from moderate dementia. He has been the subject of three other searches. In all three searches he departed from home and was found at the locations indicated on the map. The subject was reported missing by his wife at 17:30. He was last seen by her in the couple's living room. She went to prepare dinner and when she returned he was gone. He has been missing for 30 minutes. He is in average physical condition for his age. It is February, highs in the 50's F, low in the upper 30's. The yellow circles indicate the locations he was previously located.

Unit 2 – ISRID Overview

Slide 1

Unit 2: ISRID overview – Text reference chapter two page 15-30

Slide 2

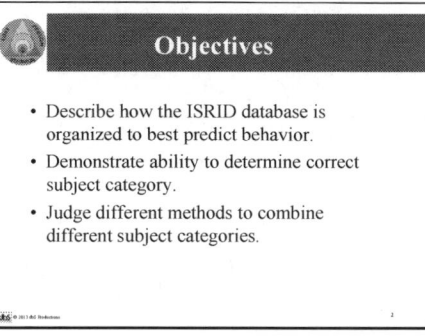

Major goals are to give an overview of ISRID, help students learn how to use the subject category hierarchy or algorithm, and the reality of the need to use multiple subject categories on some searches.

Slide 3

Lost Person Behavior

Unit 2 – ISRID Overview

Slide 4

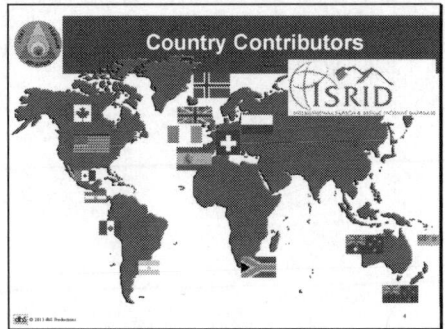

Slide 5

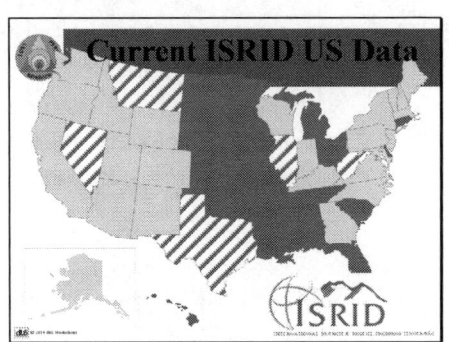

Slide current as of August 2014. Yellow means state or agency in state contributes data to ISRID. Light green mean state has indicated it will start contributing data. Dark green mean no data is collected from state.

Slide 6

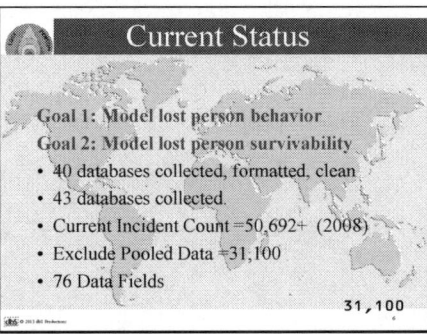

Lost Person Behavior   September 2014

Unit 2 – ISRID Overview

Slide 7

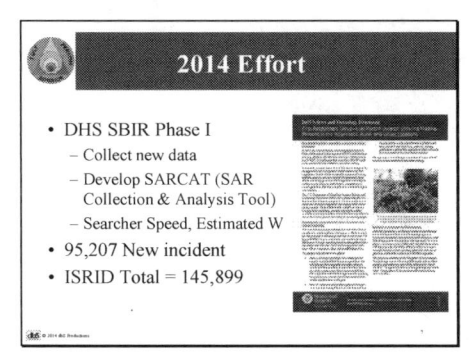

2014 Effort

- DHS SBIR Phase I
  - Collect new data
  - Develop SARCAT (SAR Collection & Analysis Tool)
  - Searcher Speed, Estimated W
- 95,207 New incident
- ISRID Total = 145,899

New data is currently being added to ISRID. Data cleaning is ongoing. New results will be reported at the end of 2014 supported by Phase II funding from DHS.

Slide 8

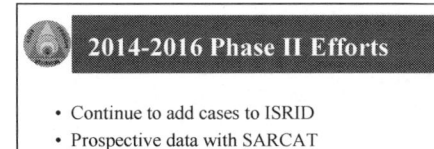

2014-2016 Phase II Efforts

- Continue to add cases to ISRID
- Prospective data with SARCAT
- New LPB Book (2015), Search Wheel
- LPB App
- Basic research: Searcher Velocity, W
- LPB/Search Management Software
- Search Management App

dbS has received a Phase II SBIR (Small Business Innovation Research) from DHS S&T. This will allow further collection of data for ISRID, the development of a data collection and Analysis Tool (SARCAT), the Second Edition of Lost Person Behavior, an updated Initial Response Search Wheel, a Lost Person Behavior App for all major operating systems, and some additional basic research. New research is focused on models to predict search velocity while searching and prospective estimated Sweep Width Values based upon vegetation, ecoregion, and slope. In addition, new models for spatial predictions will be developed (watershed, distance from destination, distance from decision point, revised LKP/PLS, point, parametric model for distance from IPP). All of this new research will be integrated search management software that will fully integrate lost person behavior and search theory.

Lost Person Behavior

Unit 2 – ISRID Overview

Slide 9

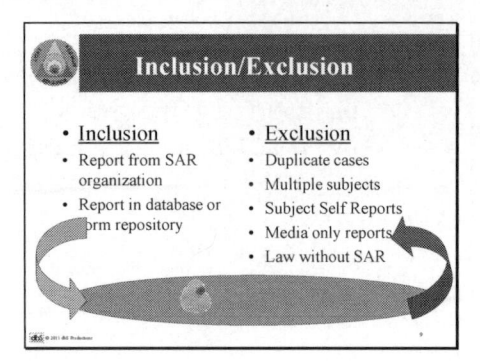

Criteria used to determine if data will go into ISRID database, or if data will be exluded

Slide 10

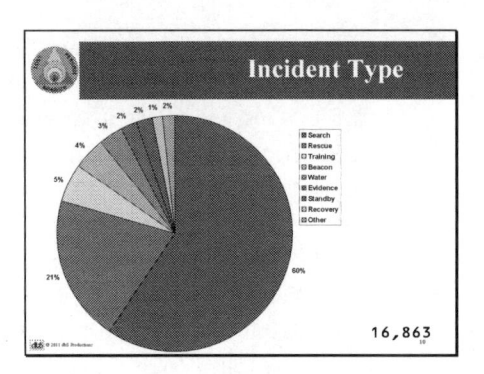

Most incidents are searches, followed by rescues. Only searches are used for most data analysis.

Slide 11

Four major types of population density terms are wilderness, rural, suburban, and urban. Wilderness and rural were combined. Suburban and Urban were combined. Urban is always labeled. If unlabeled then it is wilderness/rural data as the default.

Lost Person Behavior

Unit 2 – ISRID Overview

Slide 12

Water environment is also unique. It does not really matter if wilderness or urban if you are in or under the water.

Slide 13

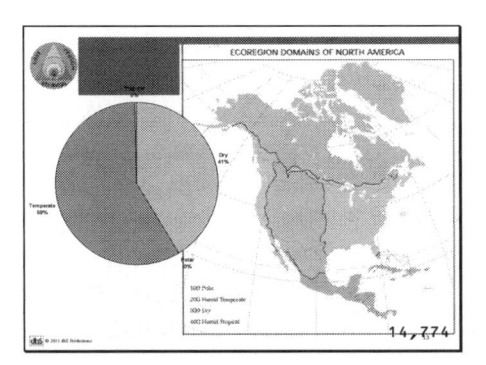

Four major EcoRegion domains of the world. Almost all data is Temperate or Dry. Polar domain should use dry and Tropical should use temperate.

Slide 14

Ecoregion is often a better predictor of subject movement than state data. Picture of temperate domain in western Washington State.

Lost Person Behavior   September 2014

Unit 2 – ISRID Overview

Slide 15

Dry domain in Eastern Washington state.

Slide 16

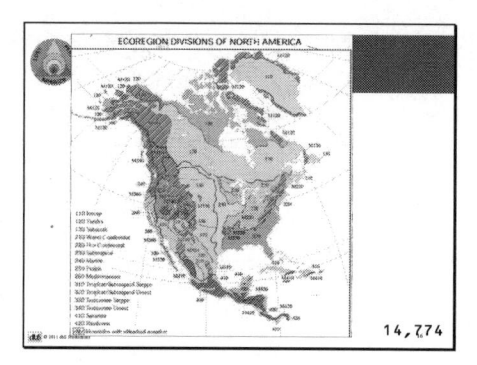

Ecoregion domains can be further broken down into divisions. At this point it is untested the role divisions play. It is hypothesized but untested that prairie (250) and Mediterranean (260) may have different results than the rest of temperate. Polar may all be the same. Also point out hash marks. Mountains may also cause different behaviors.

Slide 17

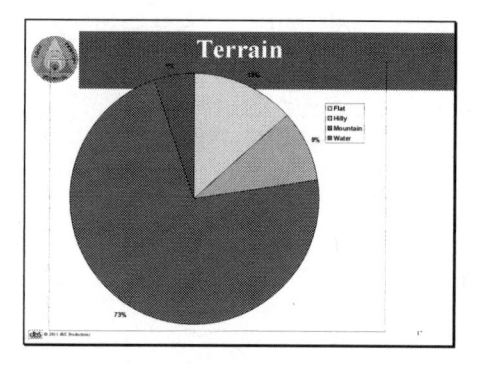

Most ISRID data comes from mountainous areas.

Lost Person Behavior

Unit 2 – ISRID Overview

Slide 18　　　　　　　　　　　　　　　　　　Example of flat (Yorktown, VA)

Slide 19　　　　　　　　　　　　　　　　　　It should be pointed out to the students that hilly terrain is combined with flat terrain.

Slide 20　　　　　　　　　　　　　　　　　　No true definition exists that defines mountainous terrain. The standard used for ISRID was a local change of approximately 1000 feet.

Lost Person Behavior

Unit 2 – ISRID Overview

Slide 21   Mt. Cook, NZ

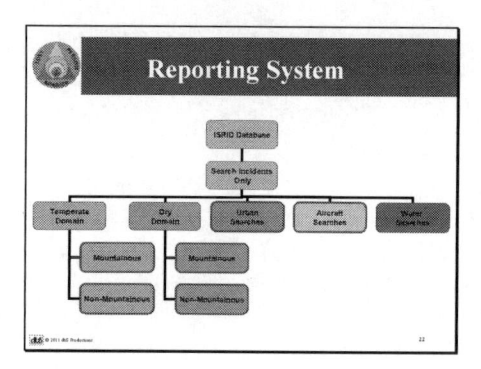

Slide 22   Overall structure of ISRID for reporting results

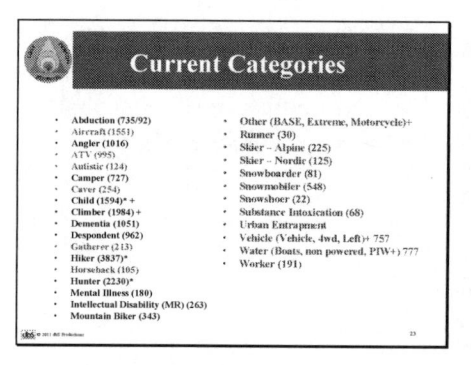

Slide 23   Current categories with incident counts. Combines searches and rescues

Lost Person Behavior   September 2014

Unit 2 – ISRID Overview

Slide 24

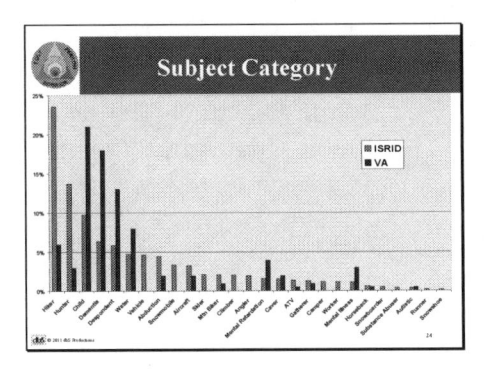

Overall distribution of subject categories. Point out regional differences may occur.

Slide 25

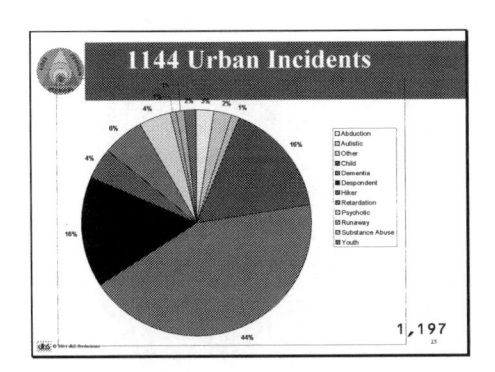

Different distributions of types of subjects for urban incidents. Dementia is the most common followed by Children and Despondent incidents.

Slide 26

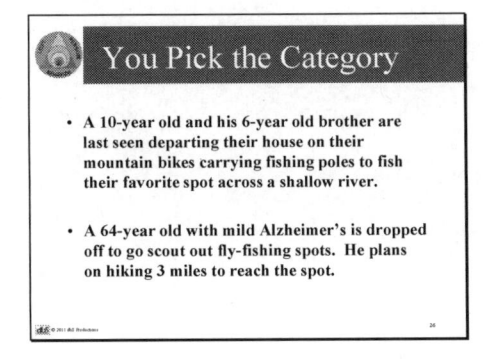

Under the older hierarchy the correct answer used to be child. Under the new algorithm the correct answer is mountain biker. However, use the slide to point out the importance of using multiple subject categories when doing real search planning. In the second incident the answer is dementia.

Lost Person Behavior — September 2014

Unit 2 – ISRID Overview

Slide 27

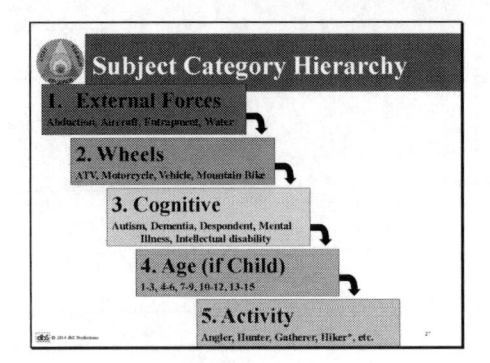

Subject Category Hierarchy
1. External Forces — Abduction, Aircraft, Entrapment, Water
2. Wheels — ATV, Motorcycle, Vehicle, Mountain Bike
3. Cognitive — Autism, Dementia, Despondent, Mental Illness, Intellectual disability
4. Age (if Child) — 1-3, 4-6, 7-9, 10-12, 13-15
5. Activity — Angler, Hunter, Gatherer, Hiker*, etc.

More detail is available for this slide on the Subject Category Algorithm or Category Wizard handout. The algorithm is based upon a hierarchy. If the highest level applies then that will be the subject category to use. In the case of finding an abandoned vehicle use the vehicle as a new IPP and then apply the hierarchy of 1-4, if none apply then the subject category would be abandoned vehicle. Hiker is starred since this is often the default category is the subject is walking. The handout is found in the forms and handout section of the Instructor's DVD. In addition, a copy of this is found in the Student Workbook.

Slide 28

Possible Matches

Subject	Primary Match	Secondary Match
Animal Search	Investigation	
Ice Skater	Nordic skier	Water-flat
Vision Quest	Gatherer	Hiker
Soldier	Worker/hiker	Extreme sport
Wheel chair	Investigation	Mountain biker

Not everything is in the book yet. Vision quest is a combination of hiker, despondent, and often mental illness. May or may not have a destination in mind.

Lost Person Behavior

Unit 2 – ISRID Overview

Slide 29

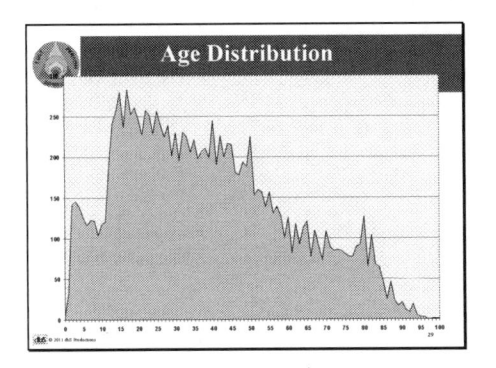

Number of incidents by age.

Slide 30

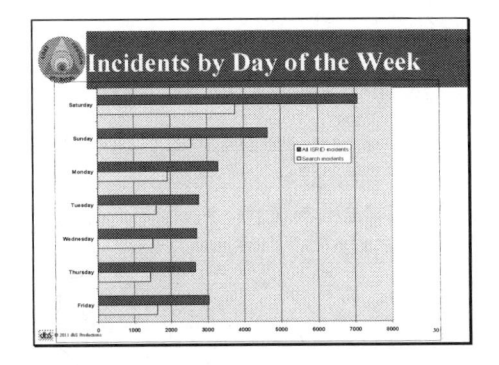

Slide 31

Most searches are short. In fact 50% (median) of searches only last 3 hours and 10 minutes. However, it is still important to be ready for the 5% (one in twenty) searches that will last more than 33 hours.

Lost Person Behavior — September 2014

Unit 2 – ISRID Overview

Slide 32

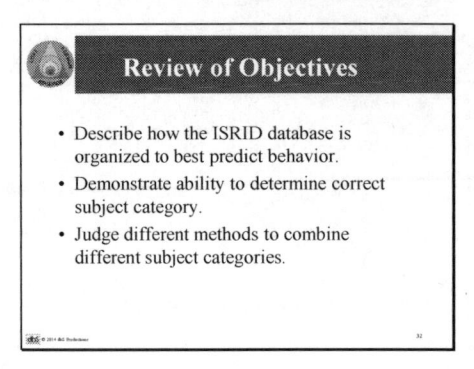

Review of Objectives

- Describe how the ISRID database is organized to best predict behavior.
- Demonstrate ability to determine correct subject category.
- Judge different methods to combine different subject categories.

Slide 33

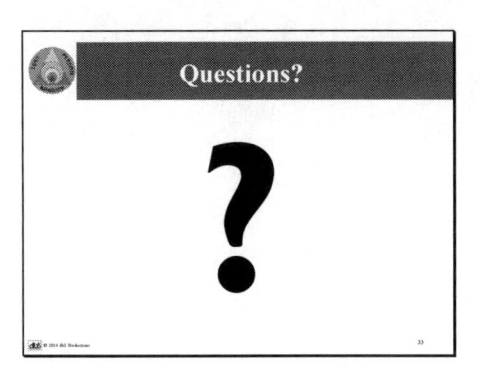

Questions?

After the questions slide, several additional slides are found. They were originally part of the LPB two day presentation. However, it was decided that this section could be shortened. However, the slides have been kept in case a presenter wants to add one or more of them back into a presentation.

Slide 34

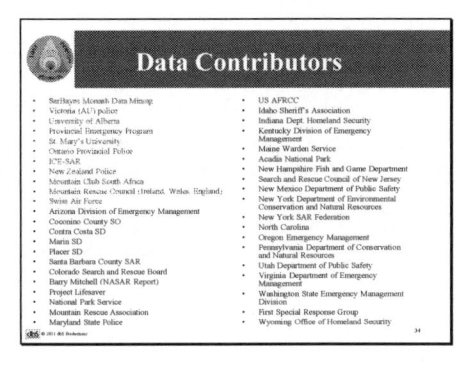

Data Contributors

Current as of March 2010

Lost Person Behavior — 42 — September 2014

Unit 2 – ISRID Overview

Slide 35

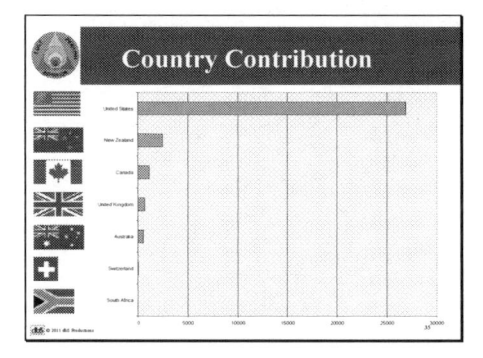

Extra slide that shows individual country contribution.

Slide 36

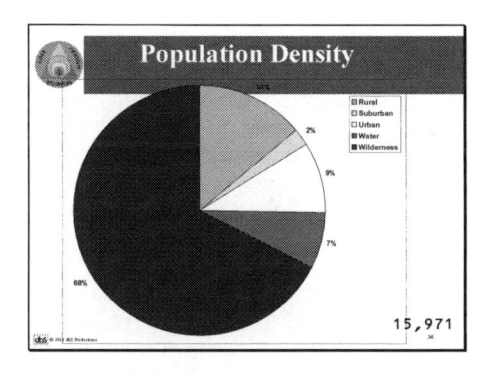

Slide 37

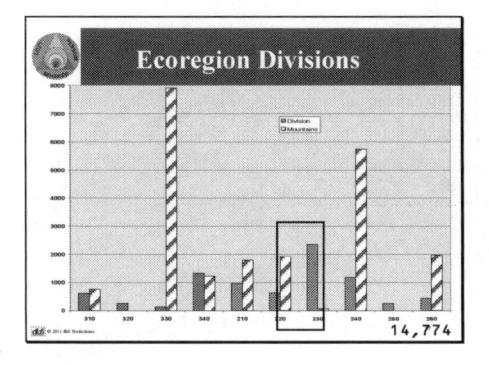

Lost Person Behavior 43 September 2014

Unit 2 – ISRID Overview

Slide 38

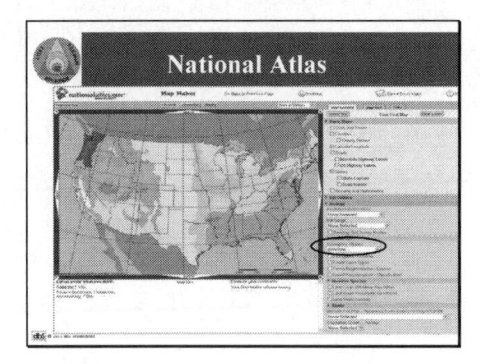

National Atlas

Slide 39

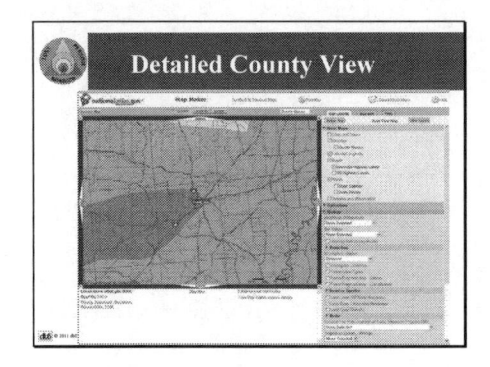

Detailed County View

Slide 40

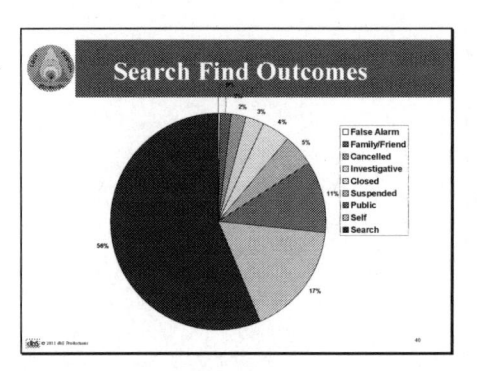

Search Find Outcomes

Lost Person Behavior

Unit 2 – ISRID Overview

Slide 41

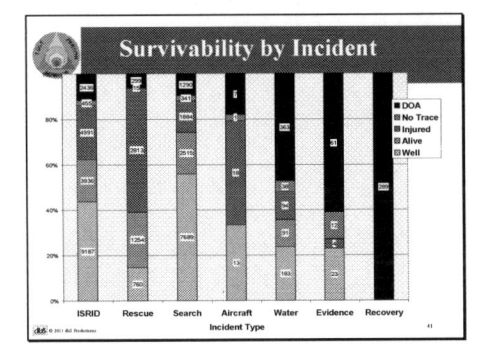

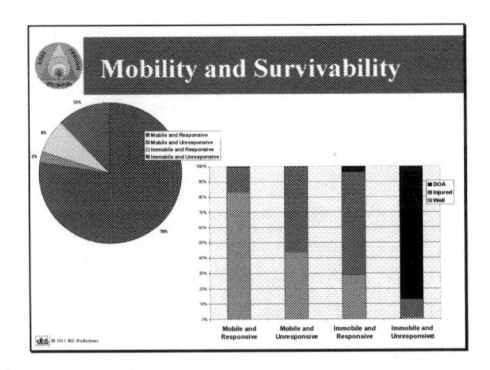

Slide 42

Slide 43

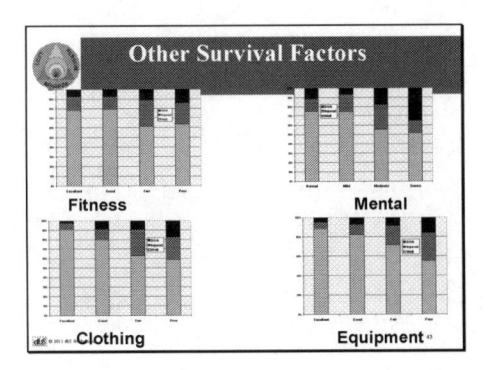

Lost Person Behavior — 45 — September 2014

Slide 44

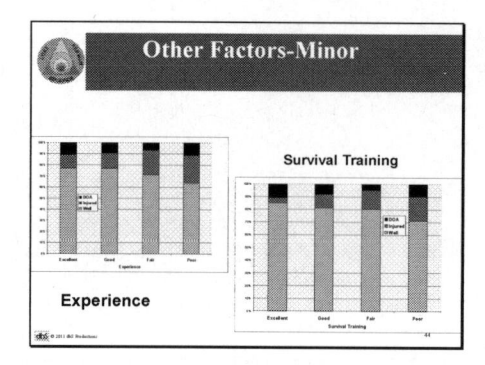

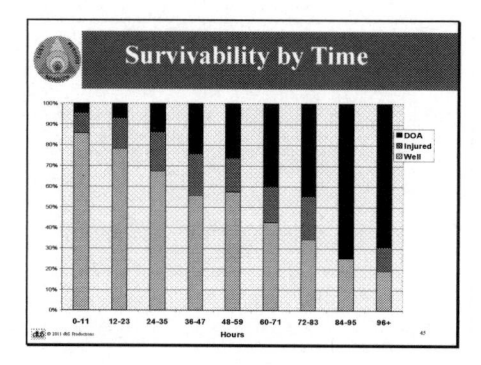

Slide 45

Unit Three – Lost Person Strategies

Slide 1

Lost Person Strategies

Text reference: Chapter 5. 53-59 and Chapter 3. Page 32-36.

Slide 2

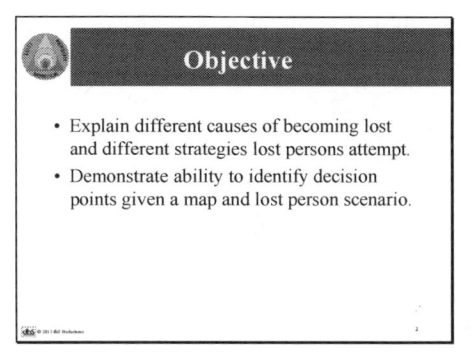

Objective
- Explain different causes of becoming lost and different strategies lost persons attempt.
- Demonstrate ability to identify decision points given a map and lost person scenario.

Two major goals of unit is to teach about various methods people use when they are lost and the role of decision points in becoming lost.

Slide 3

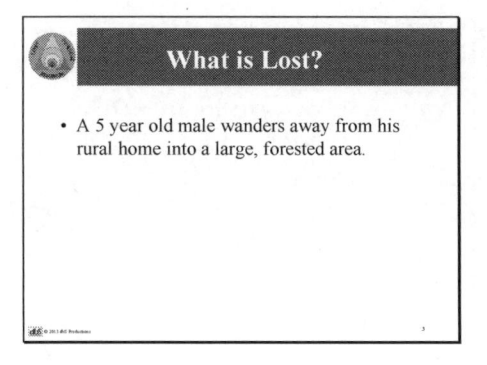

What is Lost?
- A 5 year old male wanders away from his rural home into a large, forested area.

Classic lost while other scenarios are always possible

Lost Person Behavior 47 September 2014

Unit Three – Lost Person Strategies

Slide 4

What is Lost?
- A 58 year old hiker underestimates the time it will take to reach the trail head and his wife reports him missing.

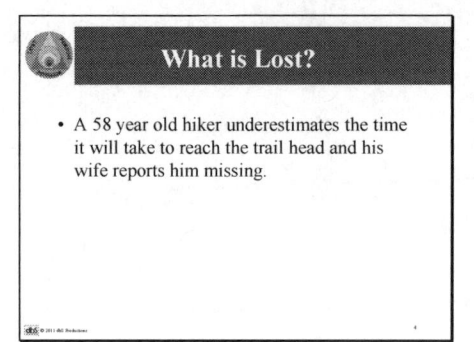

Overdue scenario

Slide 5

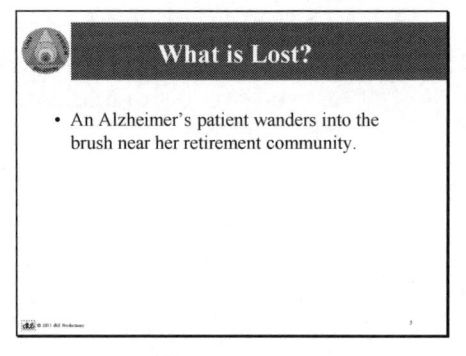

What is Lost?
- An Alzheimer's patient wanders into the brush near her retirement community.

Classic lost

Slide 6

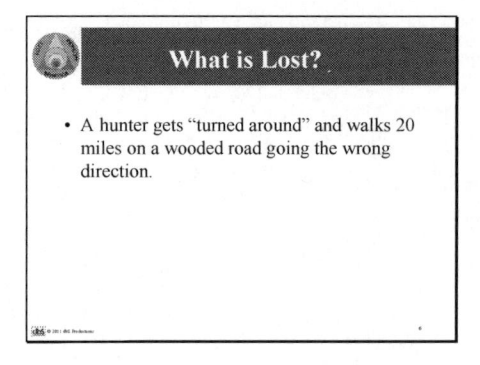

What is Lost?
- A hunter gets "turned around" and walks 20 miles on a wooded road going the wrong direction.

Lost, but note, while the subject was "lost" did not realize he was lost. So no lost strategy used, potentially no anxiety or lost reactions. Will then fall into the overdue scenario.

Unit Three – Lost Person Strategies

Slide 7

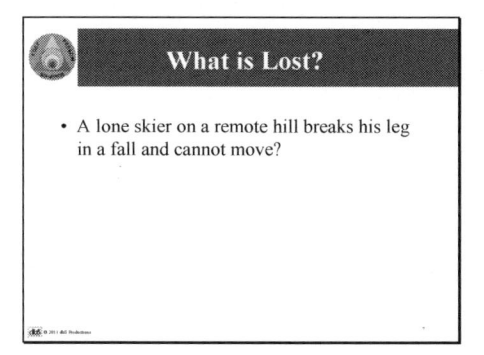

What is Lost?
- A lone skier on a remote hill breaks his leg in a fall and cannot move?

Trauma scenario. May choose to show New Zealand video for general interest. In order for the link to work an Internet connection is required.

Slide 8

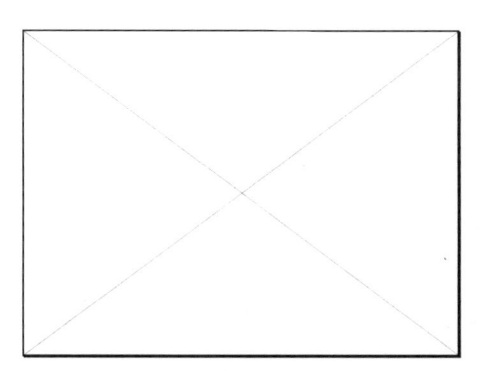

Fun little video to "demonstrate" trauma. Video clip if not internet connection is available can also be found in Video folder on Instructor DVD

Slide 9

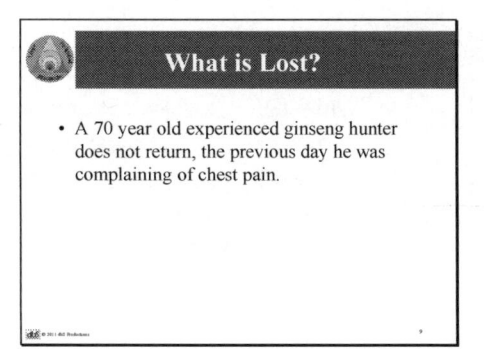

What is Lost?
- A 70 year old experienced ginseng hunter does not return, the previous day he was complaining of chest pain.

Medical scenario and medical subject category can be discussed

Lost Person Behavior

Unit Three – Lost Person Strategies

Slide 10

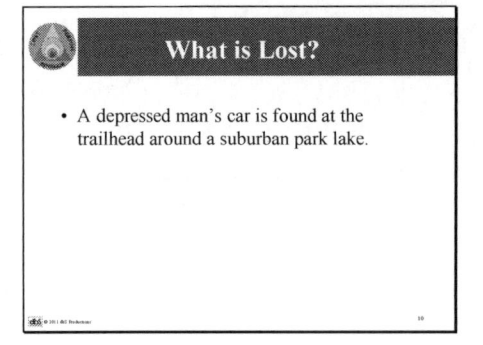

What is Lost?
- A depressed man's car is found at the trailhead around a suburban park lake.

Despondent. An example of a scenario also being the subject category

Slide 11

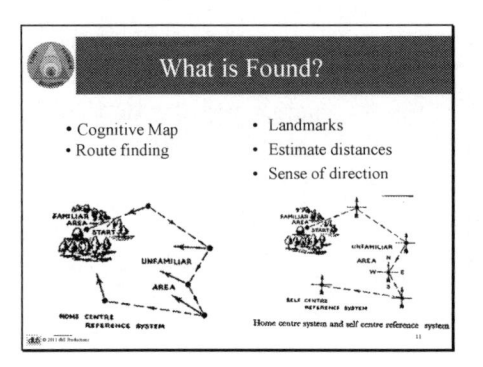

What is Found?
- Cognitive Map
- Route finding
- Landmarks
- Estimate distances
- Sense of direction

Illustrations from *Finding Your Way On Land or Sea: Reading Nature's Map* by Harold Gatty.

Slide 12

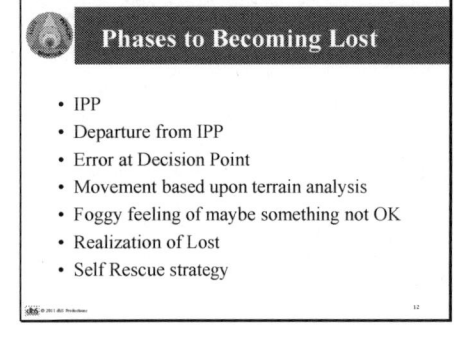

Phases to Becoming Lost
- IPP
- Departure from IPP
- Error at Decision Point
- Movement based upon terrain analysis
- Foggy feeling of maybe something not OK
- Realization of Lost
- Self Rescue strategy

Sequence of becoming lost. Important to realize that it is not until admit being lost will a self rescue strategy be developed. If errors are caught soon after the decision point it increases the chance the subject might be willing to try to backtrack.

Lost Person Behavior — September 2014

Unit Three – Lost Person Strategies

Slide 13

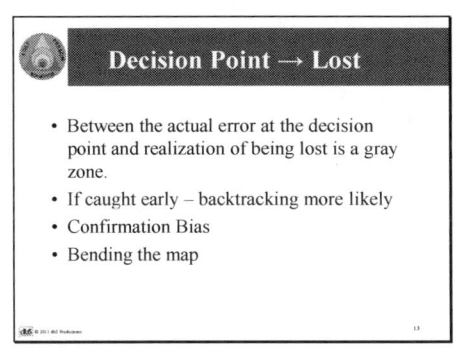

Decision Point → Lost
- Between the actual error at the decision point and realization of being lost is a gray zone.
- If caught early – backtracking more likely
- Confirmation Bias
- Bending the map

Confirmation bias is the tendency for people to favor information which supports their preconceptions or hypothesis regardless of the facts.

Slide 14

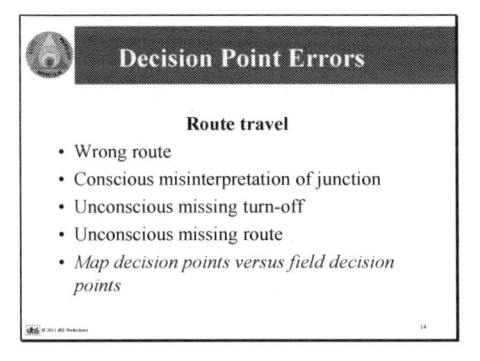

Decision Point Errors

Route travel
- Wrong route
- Conscious misinterpretation of junction
- Unconscious missing turn-off
- Unconscious missing route
- *Map decision points versus field decision points*

Decision points are discussed on this slide. Should give multiple examples of decision points. Two major types. Those that can be recognized looking at a map and those that only teams in the field could recognize. For both types it is important that field resources take some extra time to look for any sign of the subject leaving the intended path.

Slide 15

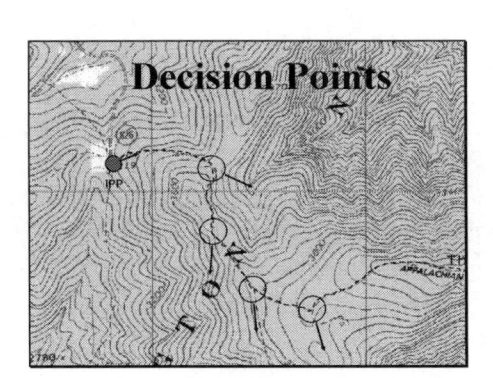

Decision Points

Search for an 17 year-old female backpacker part of a "Hoods in the woods" youth program. Entire group stopped at the IPP for lunch. The fast hikers left first, then the subject by herself, (she was observed heading the correct direction), followed by a counselor taking up the rear. At the next stopping area she was no longer present. The first decision point represents a scenario of up and over the saddle, plus a game trail most likely exists going down to the water source. In addition, a sharp turn also occurs. The second decision point is a slight change in the direction of the trail. The third decision point is a trail junction. The fourth decision point is a road that leads to a barn. It is important to realize that these are

Lost Person Behavior

Unit Three – Lost Person Strategies

mapping decision points. Points that can be recognized from a map. Team leaders should be taught to report decision points they recognize in the field but may not be on a map. In addition, when they recognize a decision point if they have some tracking or signcutting skills they should check the area carefully. The subject in this case made the mistake at decision point one.

Slide 16

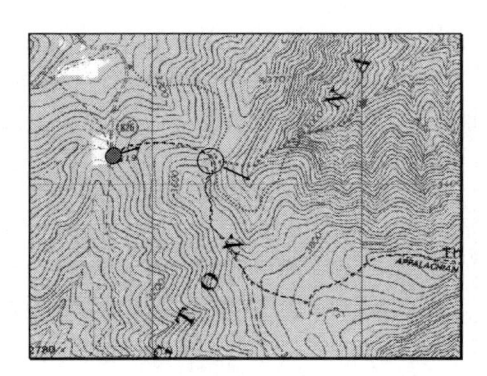

Terrain analysis. Once a mistake is made, most subjects will be guided by the terrain. The most likely path is indicated by the big blue dots. Others option include following the ridge north, (maybe to the top of the hill if the subject had a cell phone – which she did not), or maybe contouring somewhat but still winding up in the drainage.

Slide 17

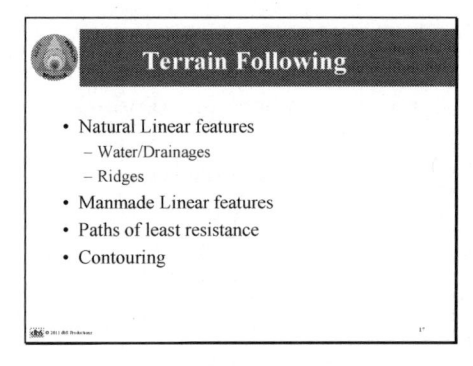

Terrain Following

- Natural Linear features
  - Water/Drainages
  - Ridges
- Manmade Linear features
- Paths of least resistance
- Contouring

Examples of how the land can push people without any real thought in certain directions. All of these are described in the book except for Contouring. Contouring typically occurs at the interface between flat terrain the subject does not want to wander into and truly steep terrain the subject does not want to climb. The only option is to stay in one place or to contour.

Lost Person Behavior

Unit Three – Lost Person Strategies

Slide 18

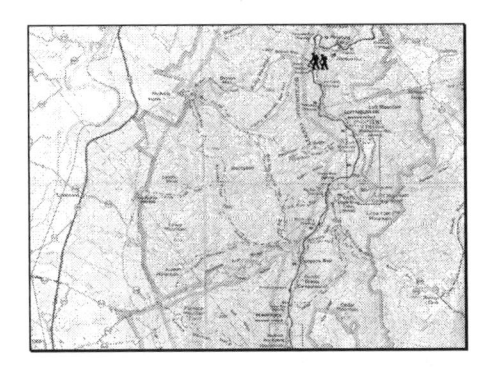

40 yo male and 34 yo female couple intended route of a spontaneous summertime afternoon hike. Intended route is to park at Brown Mountain overlook, hike west, head southwest down the Rocky Mountain Run Trail, continue downstream along the Big Run Portal Trail, then return along the Brown Mountain Trail.

Slide 19

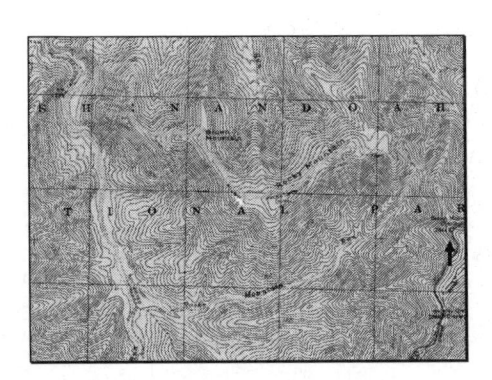

1:24,000 Topographic map showing same route. Ask students to identify potential decision points. After students point out potential decision points, click the mouse to animate the arrow. The arrow will follow the route upto the point of the decision point where the error occurred. The next slide will then show an aerial view of the decision point.

Slide 20

Google image close up of the key decision point at the intersection of the Big Run Portal trail and the Brown Mountain trail. Note bridge just after intersection (red lines are the trails).

Lost Person Behavior

Unit Three – Lost Person Strategies

Slide 21

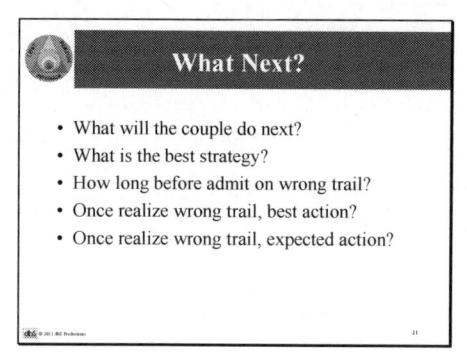

What Next?
- What will the couple do next?
- What is the best strategy?
- How long before admit on wrong trail?
- Once realize wrong trail, best action?
- Once realize wrong trail, expected action?

The instructor should lead a short class discussion on the expected actions of the couple.

Slide 22

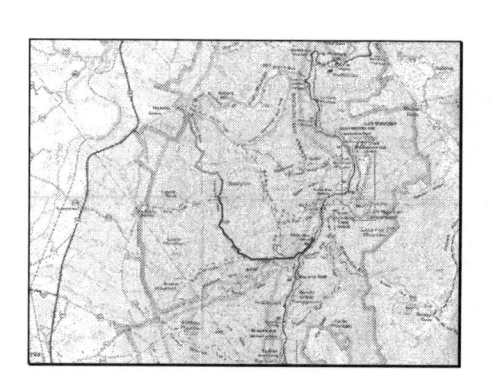

What Actually happened and the couples thoughts at the time. Hike started as planned from Brown Mountain Overlook, down Rocky Mountain Run Trail (blue), down Big Run Portal Trail (blue) but then they missed the turn off of the Brown Mountain Trail. They had focused on the bridge and got distracted. They did not have a map, but instead had attempted to memorize the map posted at the trailhead. The Big Run Portal Trail continues over the bridge and comes to a gate (yellow). The couple knew they should not leave the park, the male mentioned taking a turn to the left did not make sense, the female said try it anyhow. They continued up the Rockytop trail (orange) with a queasy feeling this was not correct. At the red segments they fully acknowledged they were on the wrong trail. After some "discussion" they decided to continue onward since it was climbing and would eventually intersect with the Skyline Drive. Upon reaching the Skyline Drive (green) they would follow it (instead of the Appalachian Trail) to get back to the car. This they did and the 10 mile hike turned into a 20 mile hike. Since they did not tell anyone where they were headed or a time back, no search was launched. However, the incident illustrates decision points, the need to follow a trail, the strong reluctance to turn around, and the near consequences

Lost Person Behavior

Unit Three – Lost Person Strategies

of being unprepared.

Slide 23

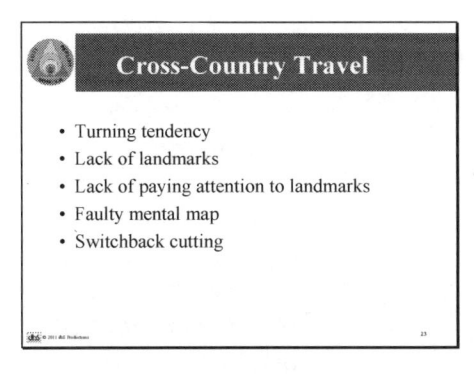

Cross-Country Travel

- Turning tendency
- Lack of landmarks
- Lack of paying attention to landmarks
- Faulty mental map
- Switchback cutting

Slide 24

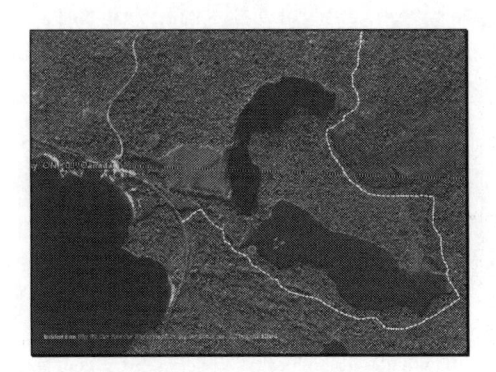

This incident is taken from the book "You are here: Why we can find out way to the moon, but get lost in the mall" by Colin Ellard. It occurred to the author on a canoe trip. It is used to illustrate how one can become lost when attempting cross-country travel and also has a decision point (two in this case). The incident takes place in Algonquin Park

Unit Three – Lost Person Strategies

Slide 25

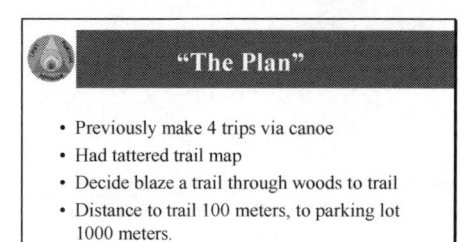

> **"The Plan"**
> - Previously make 4 trips via canoe
> - Had tattered trail map
> - Decide blaze a trail through woods to trail
> - Distance to trail 100 meters, to parking lot 1000 meters.
> - They are not 2 hours past due (expected trip 45 minutes) – What do you think happened?

Slide 26

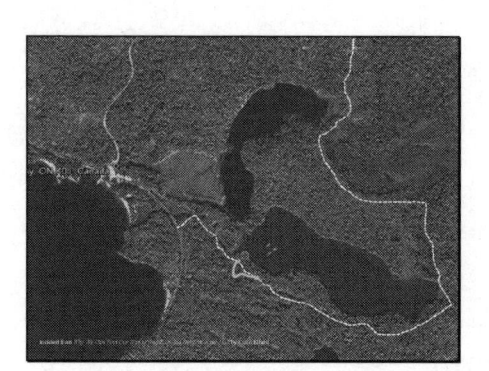

What actually happened. The subjects walked over the trail without noticing it. They natural tendency to circle caused them to come back to the trail. This time they noticed the trail and took the right turn they were expecting. However, this sent them in the wrong direction. They passed the island with the osprey's nest and unique rock they had used as a landmark before. However, this time they taught it must be associated with another lake and continued on. This case illustrates how easy it is to miss a trail when going cross-country, turning tendency, and "bending the map" or the ability to take facts and contort them to what you want to believe.

Lost Person Behavior

Unit Three – Lost Person Strategies

Slide 27

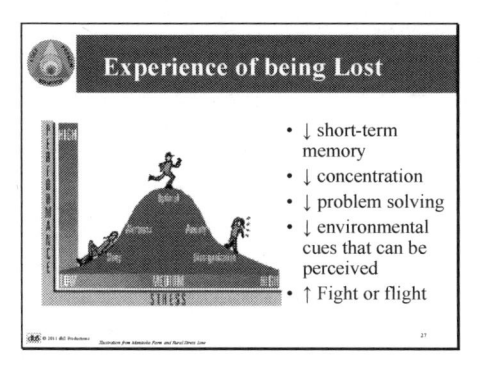

Experience of being Lost
- ↓ short-term memory
- ↓ concentration
- ↓ problem solving
- ↓ environmental cues that can be perceived
- ↑ Fight or flight

Graphic shows the Yerkes-Dodson law. Which shows an inverted U for more complex tasks. For simple tasks performance stays high with high stress. However for such cognitive demanding task such as way-finding performance will decrease when stress is high, i.e. lost.

Slide 28

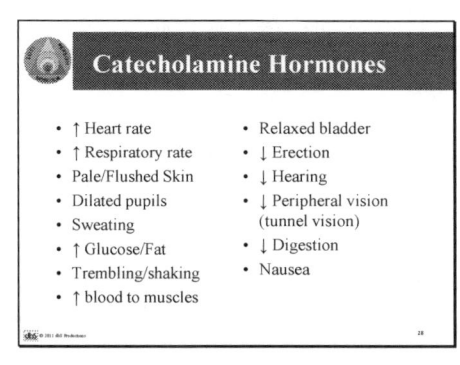

Catecholamine Hormones
- ↑ Heart rate
- ↑ Respiratory rate
- Pale/Flushed Skin
- Dilated pupils
- Sweating
- ↑ Glucose/Fat
- Trembling/shaking
- ↑ blood to muscles
- Relaxed bladder
- ↓ Erection
- ↓ Hearing
- ↓ Peripheral vision (tunnel vision)
- ↓ Digestion
- Nausea

The following is standard physiology as a result of catecholamine (adrenaline or epinephrine, norepinephrine, and dopamine) release from the adrenal glands. This is often called the flight or fight response.

Slide 29

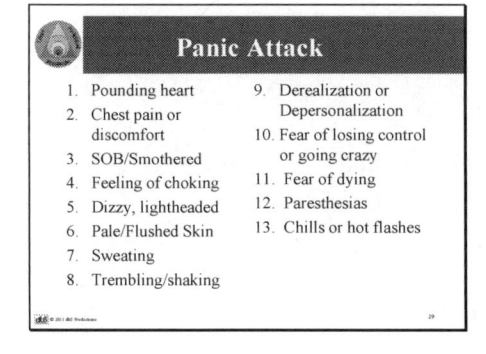

Panic Attack
1. Pounding heart
2. Chest pain or discomfort
3. SOB/Smothered
4. Feeling of choking
5. Dizzy, lightheaded
6. Pale/Flushed Skin
7. Sweating
8. Trembling/shaking
9. Derealization or Depersonalization
10. Fear of losing control or going crazy
11. Fear of dying
12. Paresthesias
13. Chills or hot flashes

The formal definition (DSM IV) of a Panic attack requires a period of intense fear or discomfort, in which four or more of the symptoms develop abruptly, reach a peak in 10 minutes, AND in the absence of real danger. Being lost in some cases may represent real danger. In other cases it may only be perceived danger. Derealization is feelings of unreality, depersonalization is being detached from oneself. Paresthesias is numbness or tingling sensations. It should be noted that many of these symptoms closely match the release of adrenalin.

Unit Three – Lost Person Strategies

Slide 30

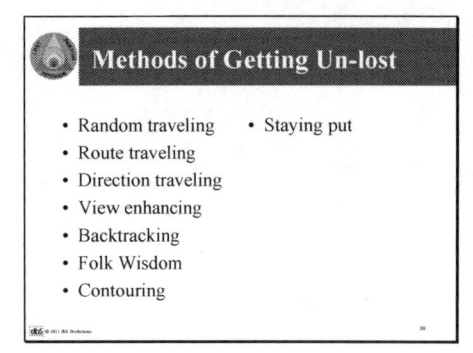

Methods of Getting Un-lost
- Random traveling
- Route traveling
- Direction traveling
- View enhancing
- Backtracking
- Folk Wisdom
- Contouring
- Staying put

List of common strategies.

Slide 31

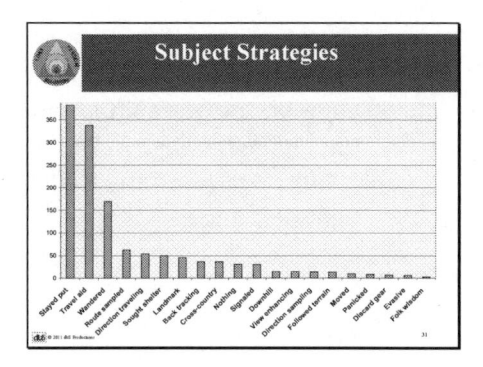

Subject Strategies

Slide 32

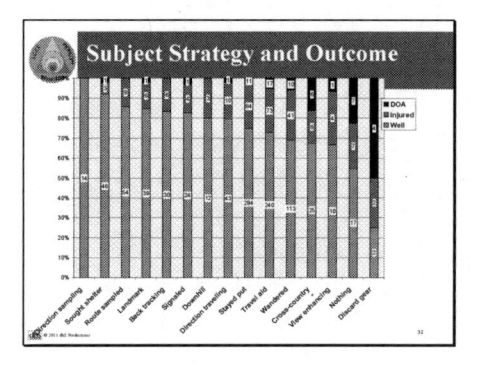

Subject Strategy and Outcome

Lost Person Behavior — September 2014

Unit Three – Lost Person Strategies

Slide 33

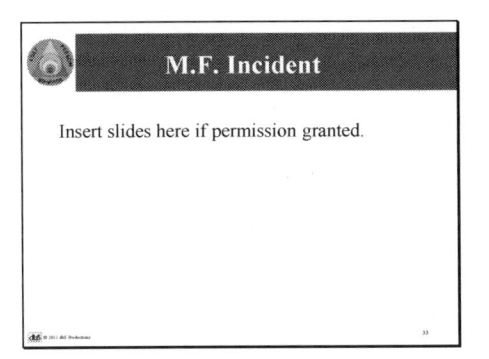

M.F. Incident

Insert slides here if permission granted.

Still working on a new presentation

Slide 34

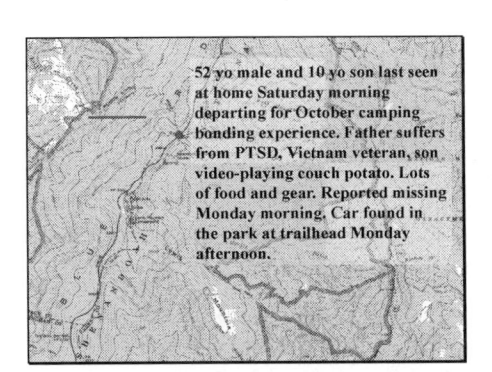

52 yo male and 10 yo son last seen at home Saturday morning departing for October camping bonding experience. Father suffers from PTSD, Vietnam veteran, son video-playing couch potato. Lots of food and gear. Reported missing Monday morning, Car found in the park at trailhead Monday afternoon.

The IPP is the parking lot on Skyline Drive at Bearfence mile. The intended route was to hike north along the Appalachian Trail, East down the Conway River Road, and then camp along the Conway River. On Sunday, they planned to pick up the Slaughter Road west back up Lewis Mountain and end back at their car. The purple dots show the first days effort. The red dot shows their campsite and revised PLS based upon witnesses who spoke with them. The orange dots represents a short cut described to them by the witnesses and the yellow dots represents the path they should have followed to return back to their car.

Lost Person Behavior

Unit Three – Lost Person Strategies

Slide 35

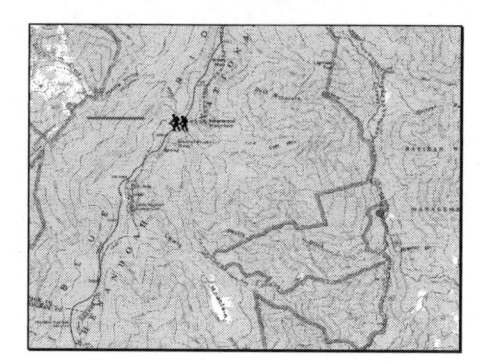

The IPP is the parking lot on Skyline Drive at Bearfence mile. The intended route was to hike north along the Appalachian Trail, East down the Conway River Road, and then camp along the Conway River. On Sunday, they planned to pick up the Slaughter Road west back up Lewis Mountain and end back at their car. The purple dots show the first days effort. The red dot shows their campsite and revised PLS based upon witnesses who spoke with them. The orange dots represents a short cut described to them by the witnesses and the yellow dots represents the path they should have followed to return back to their car.

Slide 36

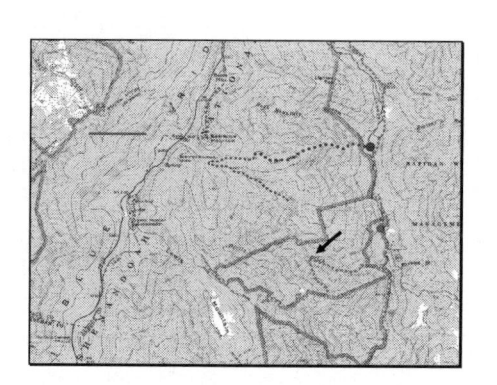

After making the error at the decision point in Devils Ditch indicated by the black arrow they followed the linear feature of a secondary trail up. The secondary unmarked trail ends and the continued to follow the linear feature of the drainage up. The camped Sunday night. On the following day (color dark blue) the continued up some more following the drainage but when it ended they switched strategies to one of contouring the mountain. This was partly due to being out of shape. The vegetation was thick and they made little progress. The slept Monday night near the summit of Clift Mountain. On Tuesday, they changed strategies again and decided to head back down the mountain to try and find the fire road again. They had descended the mountain and had reached a small stream (but no road) and were contemplating going back up the mountain to try and find Skyline Drive when two search teams were having an argument nearby. The father and son shouted and the teams connected with them. They stated they were never lost!

Lost Person Behavior 60 September 2014

Unit Three – Lost Person Strategies

Slide 37

See Activity Guide 3-6 for a full description of incident. Point out car and camp. Show how picture depicts the camp area.

Slide 38

Black lines show the path the subject reports was his route. The numbers represent where he spent each night. It is belived he slept for 24 hours on nights 2 and three. He eventually walked out.

Slide 39

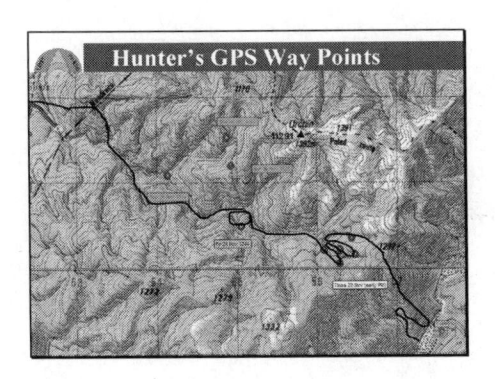

The black line shows where the subject thought he went. However, since the subject carried a GPS and recorded several waypoints it is possible to determine where he actually traveled. The actual route starts the same, he does have a day of missing time and it is thought he might have slept for 24 hours of so on Saturday. The blue line shows where he actually went. A close examination will show he thought he came down a drainage, but it was the wrong drainage. The actual route was confirmed by finding physical clues by trackers.

Unit Three – Lost Person Strategies

Slide 40

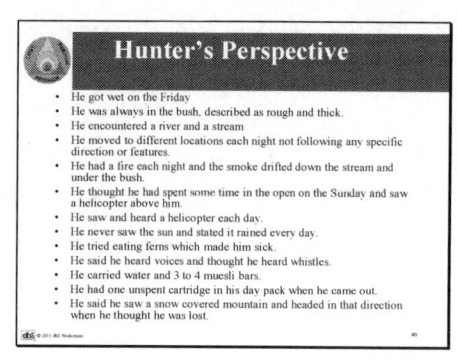

Hunter's Perspective
- He got wet on the Friday
- He was always in the bush, described as rough and thick.
- He encountered a river and a stream
- He moved to different locations each night not following any specific direction or features.
- He had a fire each night and the smoke drifted down the stream and under the bush.
- He thought he had spent some time in the open on the Sunday and saw a helicopter above him.
- He saw and heard a helicopter each day.
- He never saw the sun and stated it rained every day.
- He tried eating ferns which made him sick.
- He said he heard voices and thought he heard whistles.
- He carried water and 3 to 4 muesli bars.
- He had one unspent cartridge in his day pack when he came out.
- He said he saw a snow covered mountain and headed in that direction when he thought he was lost.

Slide 41

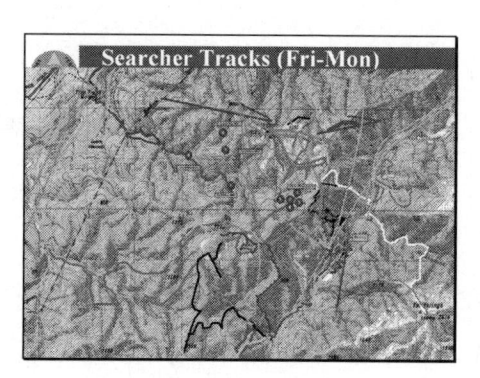

Searcher Tracks (Fri-Mon)

Searcher tracks shown. Helicopter tracks are silver and orange/tan. It can be seen a helicopter covered his route and a search team (red)
covered his route. This was the same search team that left a message for him and whose voices he most likely heard.

Slide 42

Knight Incident
- 41-year-old visually impaired male "solo" hiking AT south as part of a group.
- 21 Apr – Started hiking at MP 13.1 south
- 26 Apr – IPP Punchbowl shelter MP 51.6
- 26 Apr – Destination Johns Hollow Shelter
- 28 Apr – Final Destination MP 76.1
- 29 Apr – Scheduled Flight back to Detroit

This is activity 3-7 in the IG Activity Guide. Legally blind, could see out of one eye, could read something about 2 inches in front of his face, and had no depth perception.

Unit Three – Lost Person Strategies

Slide 43

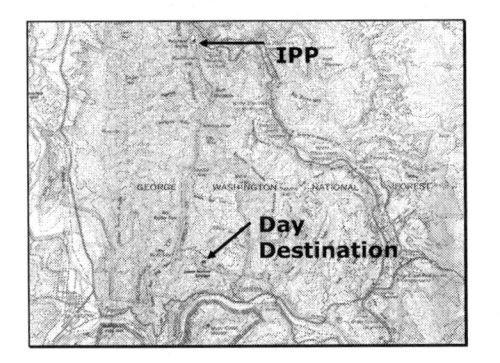

Subject departed Punchbowl Shelter with plans to hike to Johns Hollow Shelter that day. He never arrived.

Slide 44

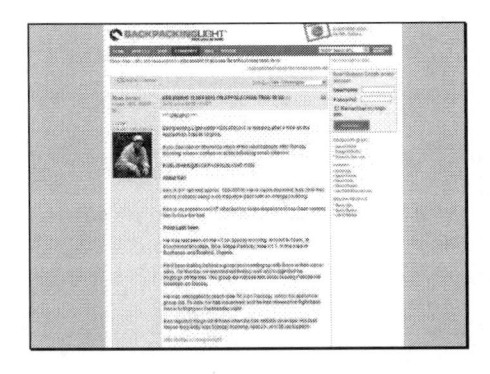

Slide 45

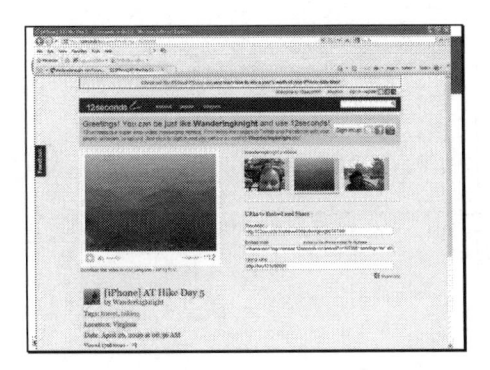

The hyperlink no longer works. The message he left stated he was hiking, he was feeling tired, the trail was more difficult than he thought. Time stamp caused confusion since shadows did not match. Eventually learned time stamp was PST while actual search area was EST.

Unit Three – Lost Person Strategies

Slide 46

Knight Information
- The hiker is Ken Knight, a visually impaired hiker. Ken is also the voice on podcasts from Backpackinglight.com
- Ken is 5'4" and 41-years-old. The picture was taken on the hike. He has a blaze orange backpack, tent, sleeping bag, and "minimal food". He was dressed appropriately for the conditions and water sources were widely available

Hiking companions had taken a picture of subject during hike. So picture is of actual clothing and current appearance.

Slide 47

- IPP Punchbowl Shelter
- Destination Johns Hollow Shelter
- Updated PLS South of Bluff Mtn
- Direction of Travel - Southbound

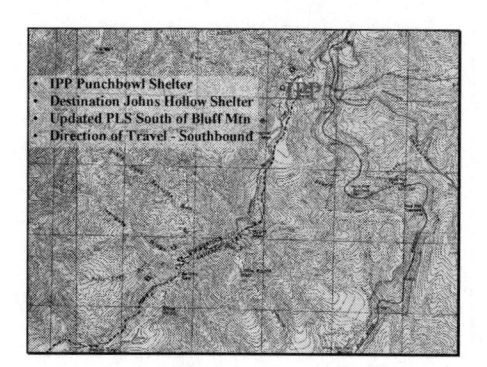

Map shows both IPP and revised PLS which had him continuing his hike to the south.

Slide 48

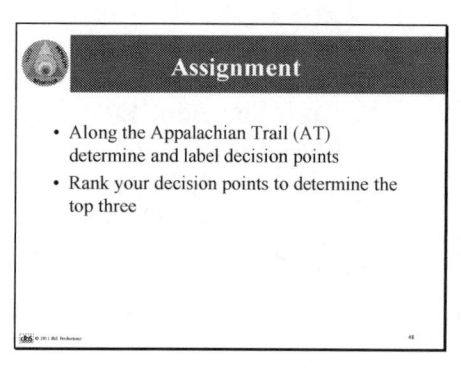

Assignment
- Along the Appalachian Trail (AT) determine and label decision points
- Rank your decision points to determine the top three

Unit Three – Lost Person Strategies

Slide 49

Use notes from IG manual to point out Find Location.

Slide 50

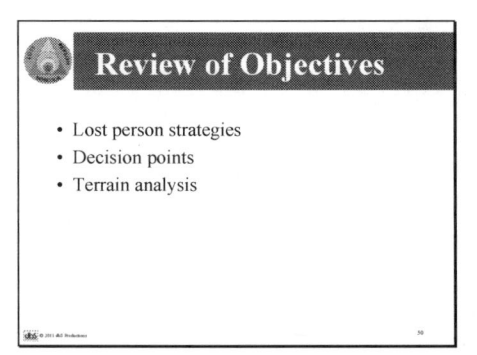

Review of Objectives
- Lost person strategies
- Decision points
- Terrain analysis

Review key points of the unit.

Slide 51

Questions

?

Lost Person Behavior   September 2014

Unit Three – Lost Person Strategies

Unit 4 – Myths and Legends

Slide 1

Text reference chapter 6. Page 61-73 Full story on hoax picture at http://www.hoax-slayer.com/shark-helicopter-photo.shtml. Fact or Fiction displays on second click.

Slide 2

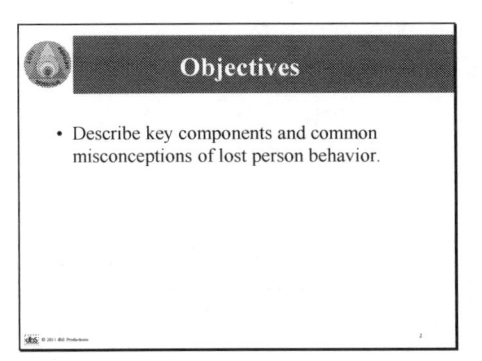

Goal of this presentation was to cover several different topics in an interactive fashion. Students should be instructed to take out the pre-test.

Slide 3

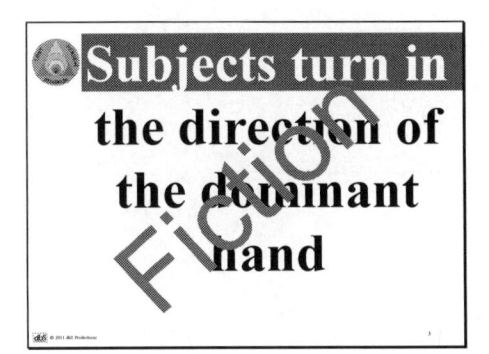

This is a common belief, but in an operational sense, the dominant hand does not predict.

Unit 4 – Myths and Legends

Slide 4

Dominant Hand - False
- Syrotuck – no
- Koester – no
- Silverstein and Salamons – no
- Some basis of myth

Specific research from the perspective of lost people in regards to dominant hand predicting turning

Slide 5

Turning Tendencies
Left-Handed: 70%
Right-Handed: 69%, 47%

See LPB textbook for full explanation. Testing done in library. Students could turn left or right at end of hall. In control conditions dominant hand appears to influence in the US. In the UK where driving changes a strong right tendency, different results occur. In real world too many factors influence, wind direction, uphill vs downhill, different weight on shoulder straps, thick brush vs easy way.

Slide 6

Lost Subjects can Walk in Circles

They most certainly can

Lost Person Behavior   68   September 2014

Unit 4 – Myths and Legends

Slide 7

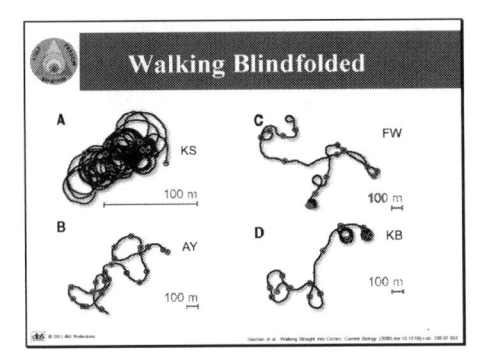

Walking in circles is true. Shown are tracks from four different participants who were blindfolded and asked to walk a straight line. They walked for a total of 50 minutes in a large field at an airstrip. The citation for the research is found on the slide. This is an excellent study that instructor's should read for background. Point out both clockwise and counterclockwise circling occurs by the same person. This refutes a structural differences (as originally hypothesized in the text Lost Person Behavior). In addition, the researcher built shoes with a slight lift to one leg and that did not cause circling in any one particular direction. The current hypothesis is that humans are not good at dead reckoning and need frequent resets to the internal compass.

Slide 8

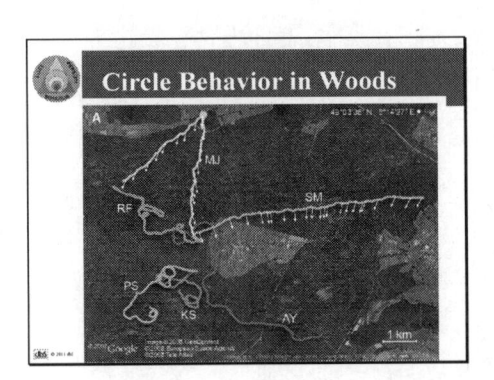

Six different subjects asked to walk a straight line through a hardwood forest. The subjects with the yellow lines (MJ and SM) walked while the sun was shining (the white arrows point to where the sun was in the sky). MJ was redirected back into the woods when he reached the end of the woods. The blue lines are subjects who walked during cloudy conditions. They were never aware of walking over their path. Subjects were not blindfolded.

Lost Person Behavior   September 2014

Unit 4 – Myths and Legends

Slide 9

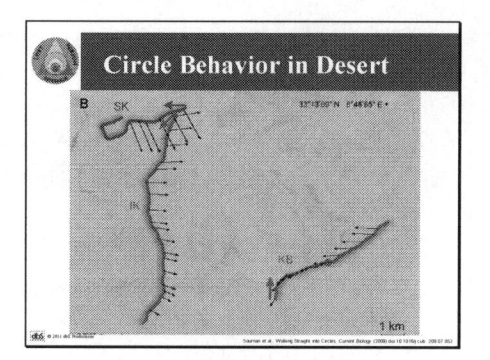

Same experiment, but conducted in the desert. Red indicates the sun was shining and blue was done during the night. A full moon was present except for at the end when it went behind a cloud.

Slide 10

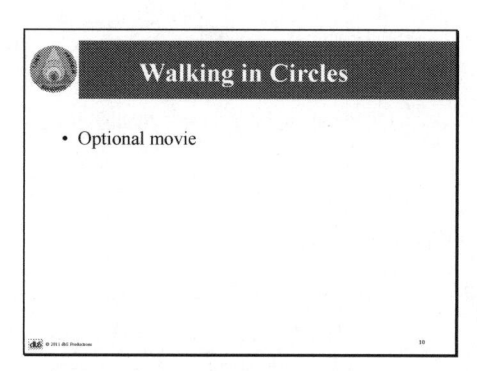

Optional movie is found on Instructor's DVD.

Slide 11

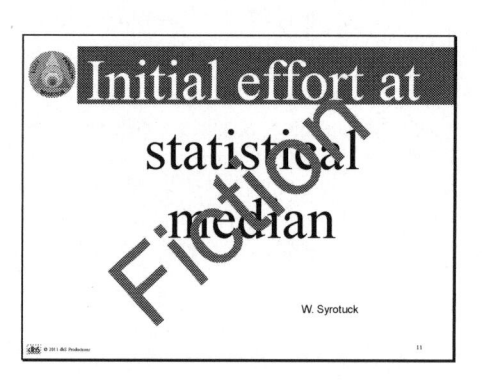

Lost Person Behavior   70   September 2014

Unit 4 – Myths and Legends

Slide 12

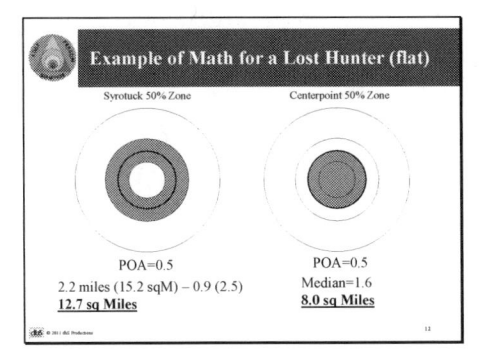

Original Syrotuck teaching suggested searching between the 25% and 75% ring first. Math with probability density(Pden) clearly shows it makes more sense to search between the IPP – 50% first.

Slide 13

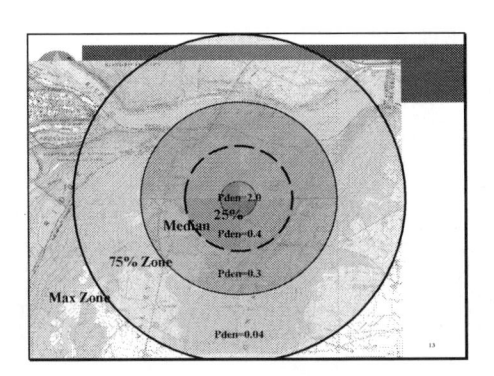

Figures shows Pden values for each of the quartile rings. Fairly typical 10 fold drop between 25% and 50%, and another ten-fold drop between Inside 75% and 95%

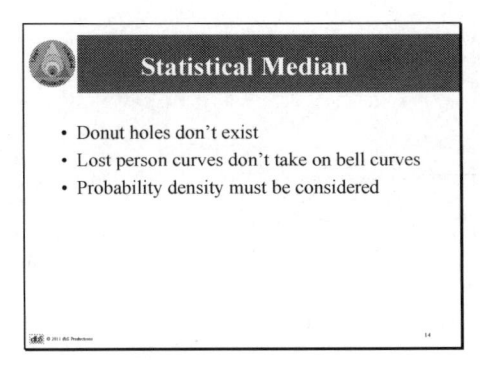

Slide 14

Statistical Median

- Donut holes don't exist
- Lost person curves don't take on bell curves
- Probability density must be considered

Unit 4 – Myths and Legends

Slide 15

[Slide: "Lost Persons go downhill (at least 90%)" marked "Fiction"]

Slide 16

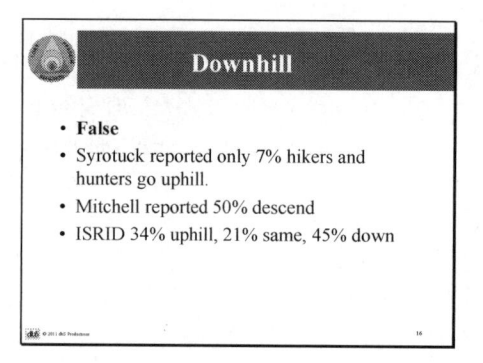

[Slide: Downhill
- False
- Syrotuck reported only 7% hikers and hunters go uphill.
- Mitchell reported 50% descend
- ISRID 34% uphill, 21% same, 45% down]

Syrotuck only reported uphill values. Mitchell data is also known as the NASAR study and reported uphill and downhill.

Slide 17

[Slide: "Profiles tell you the subject's location" marked "Fiction"]

Profiles only give relative probabilities and generalities.

Unit 4 – Myths and Legends

Slide 18

Slide 19

Taken from Barry Mitchel's NASAR study

Slide 20

Investigation especially initial investigation that has not been verified by multiple sources has often proven to be incorrect. A search planner should always send resources to the statistically most likely areas, even if investigation suggests otherwise. The priority of tasking will still be set by investigation and looking at all factors.

Unit 4 – Myths and Legends

Slide 21

One study suggested stopping searching after 51 hours. Most searchers know this is not correct. The basis of this statement was a paper by Annette Adams et al. "Search is a Time-Critical Event: When Search and Rescue Missions May Become Futile: in Wilderness Environ Med 2007;18:95-101. Fifty-one (51) hours represents the point where 99% of survivors had been found. This is similar to statistics collected by ISRID.

Slide 22

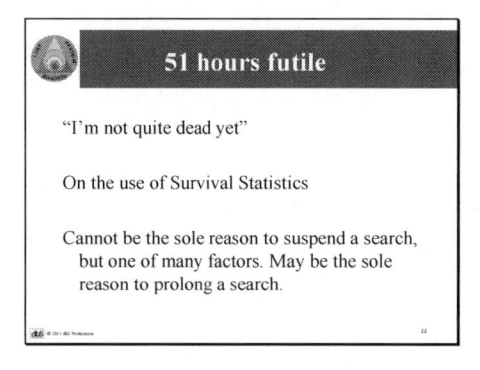

Slide 23

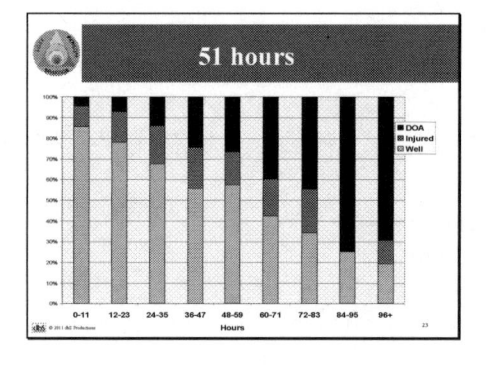

Looking at the 48-59 hour bin (2 days – 2.5 days) it can be seen that 75% of the subjects found during this time frame (after being last seen) are found alive. Clearly 51 hours cannot be used as a rule of thumb for suspending search incidents.

Lost Person Behavior

Unit 4 – Myths and Legends

Slide 24

Subjects Travel Further in Mountainous Terrain

Depends

The answer depends because it is dependent somewhat on how students interpret the question. In classic Syrotuck studies he showed that subjects in temperate areas traveled further from the IPP than in temperate flat areas. This result is supported by ISRID. However, in the dry domain no difference is found. In addition, for those with dementia, they travel further in flat terrain than in mountainous terrain. So it does depend upon the subject category as well.

Slide 25

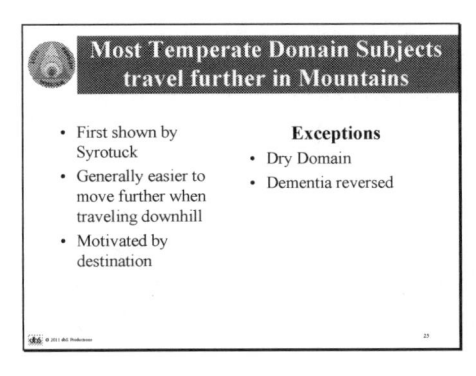

Most Temperate Domain Subjects travel further in Mountains

- First shown by Syrotuck
- Generally easier to move further when traveling downhill
- Motivated by destination

Exceptions
- Dry Domain
- Dementia reversed

Slide 26

The longer the search the further away lost subjects are found

Fiction

Lost Person Behavior

Unit 4 – Myths and Legends

Slide 27

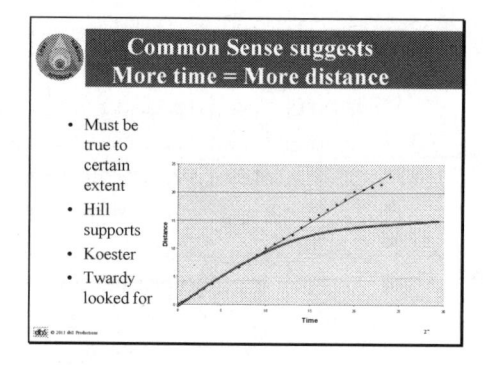

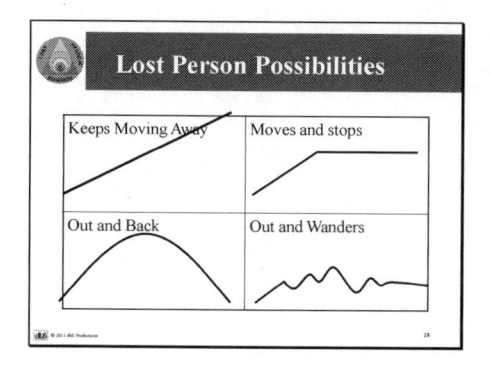

Slide 28

Slide 29

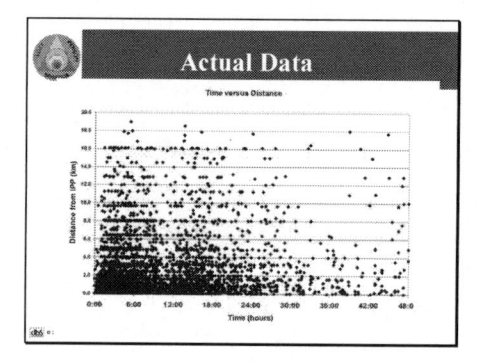

Actual data from ISRID, after excluding statistical outliers, and wheeled subject categories. No predictive pattern emerges. However, the tendancy to round numbers off to whole values can be seen, and a little null at 12 hours is possible (shift change less likely to make a find).

Lost Person Behavior

Unit 4 – Myths and Legends

Slide 30

Basis of using ecoregions

Slide 31

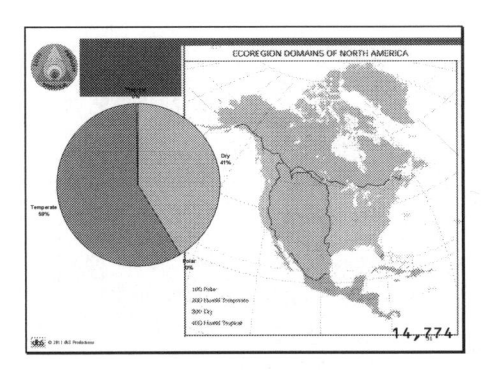

Once again showing the four major Ecoregion domains. If class is not in North America should switch out the illustration with maps that are available for Australia/New Zealand, or Europe. It can also be seen that very little Tropical or Polar data exists. Current work suggests that tropical should use temperate and Polar should use dry.

Slide 32

(Left) Temperate area of Washington State (Right) Dry area of Washington State (valley irrigated for orchards). State data may not always point the correct statistical picture. Important to use the Ecoregions.

Lost Person Behavior

Unit 4 – Myths and Legends

Slide 33

We can take the Search out of Search and Rescue — *Fiction*

Title of a NASA report and several other reports.

Slide 34

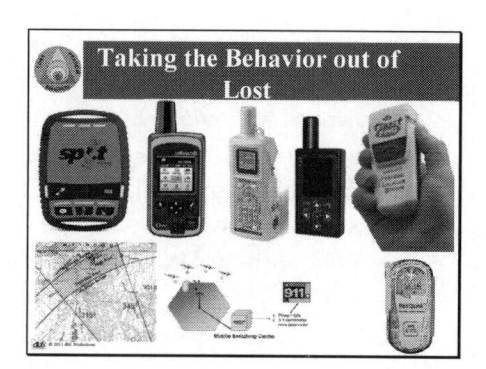

Taking the Behavior out of Lost

Examples of SEND devices (Satellite Emergency Notification Device). Spot using globalstar systems which is one way. InReach, Solara and Shoutnano which use Iridium network and allow two-way communication. The ACR SARlink view which is a 406 PLB with the ability to see GPS coordinates and send an I'm ok message. The McMurdo Fast Find PLB is also shown. Cell phone forensics is now offered by the AFRCC for SAR incidents. This is a technical topic, Appendix B in *Lost Person Behavior* covers it in more detail.

Slide 35

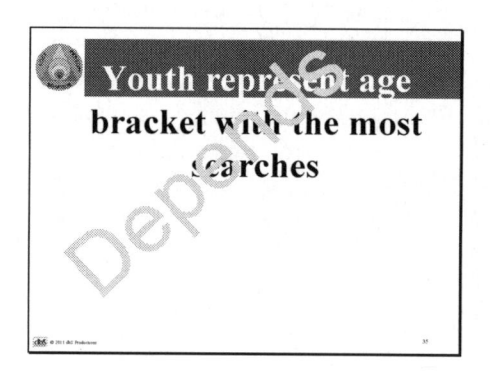

Youth represent age bracket with the most searches — *Depends*

Lost Person Behavior — 78 — September 2014

Unit 4 – Myths and Legends

Slide 36

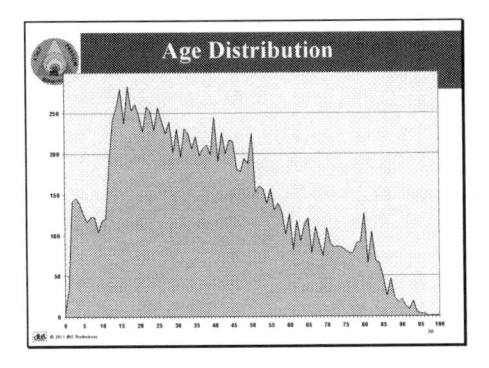

While youth account for the mode (largest number for any particular year), far more incidents occur for adults.

Slide 37

Slide 38

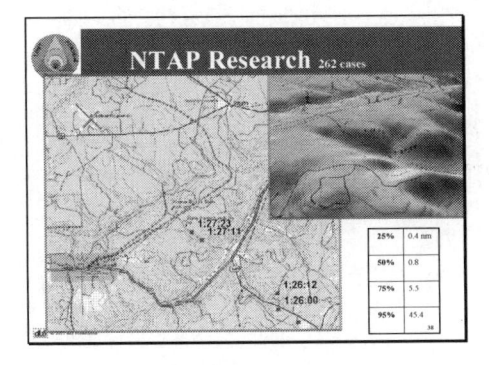

Radar data simply gives a new IPP. However, in some cases (VFR flights when squaking 1200) it may even be for the wrong aircraft.

Unit 4 – Myths and Legends

Slide 39

> **PLB's are used for land SAR and will have a lower false alarm rate than ELTs**
>
> *Fiction*

Slide 40

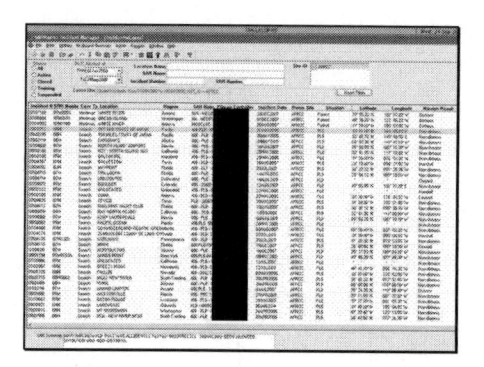

PLBs still have a high non-distress rate. Data shows only one distress out of the entire screenshot.

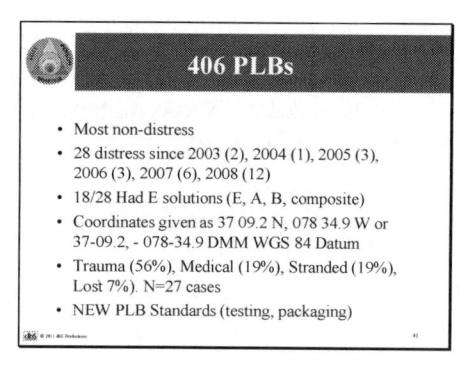

Slide 41

406 PLBs

- Most non-distress
- 28 distress since 2003 (2), 2004 (1), 2005 (3), 2006 (3), 2007 (6), 2008 (12)
- 18/28 Had E solutions (E, A, B, composite)
- Coordinates given as 37 09.2 N, 078 34.9 W or 37-09.2, -078-34.9 DMM WGS 84 Datum
- Trauma (56%), Medical (19%), Stranded (19%), Lost 7%). N=27 cases
- NEW PLB Standards (testing, packaging)

Unit 4 – Myths and Legends

Slide 42

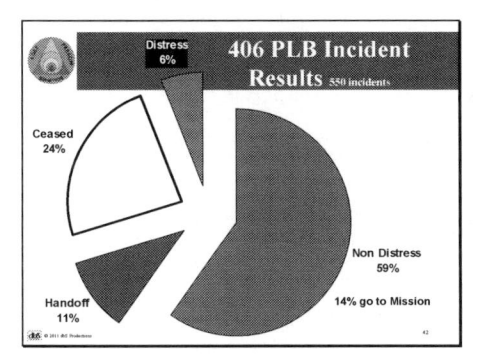

Most non-distress PLB alerts are solved with a phone call. The importance of registering the device.

Slide 43

Slide 44

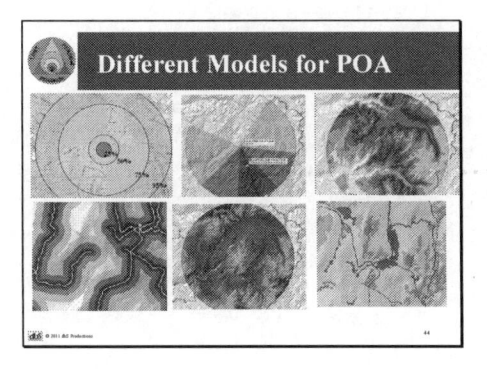

View of the six models found in Lost Person Behavior. From Top, Left to Right; Ring, dispersion, elevation, track-offset, mobility, and find location. These will be discussed in more detail in another section.

Unit 4 – Myths and Legends

Slide 45

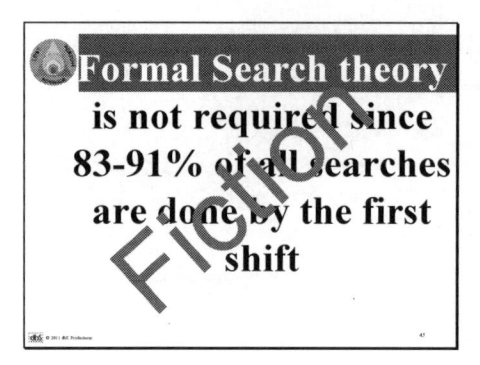

> **Formal Search theory is not required since 83-91% of all searches are done by the first shift**
>
> *Fiction*

Slide 46

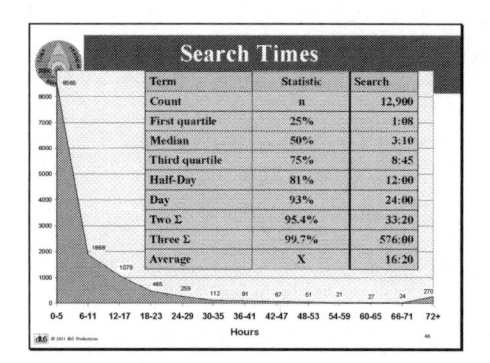

Term	Statistic	Search
Count	n	12,900
First quartile	25%	1:08
Median	50%	3:10
Third quartile	75%	8:45
Half-Day	81%	12:00
Day	93%	24:00
Two Σ	95.4%	33:20
Three Σ	99.7%	576:00
Average	X	16:20

Same search time graph shown before. Most searches end quickly. However, it is certain that some will become multi-operational period. Therefore, while this class does not cover formal search theory it is still critical.

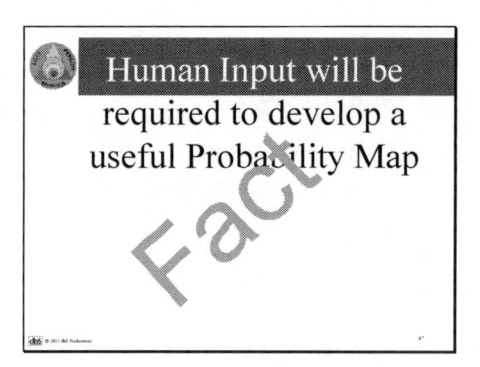

Slide 47

> **Human Input will be required to develop a useful Probability Map**
>
> *Fact*

Lost Person Behavior

Unit 4 – Myths and Legends

Slide 48

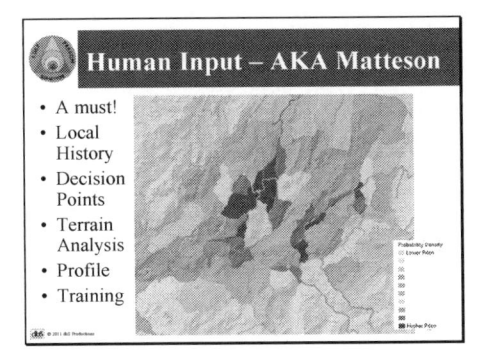

Research in Matteson consensus shows it typically does a little bit better than a computer generated model – if the correct people are doing the consensus.

Slide 49

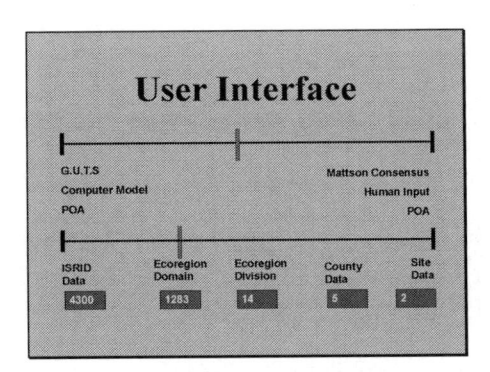

One potential user interface would allow the user to determine how much weight to give the Mattson over the computer models. It is also possible to weight local data over international data.

Slide 50

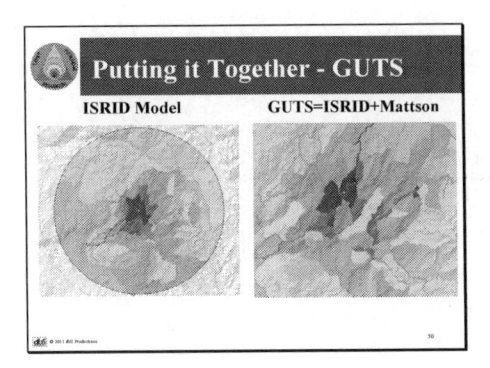

The early research also shows that the best model usually is a combination of both the computer generated model plus the consensus. The two models are similar but slightly different. Model in the illustration was generated with ESRI software.

Lost Person Behavior

Unit 4 – Myths and Legends

Slide 51

Review of Objectives
- Describe key components and common misconceptions of lost person behavior.

Slide 52

Questions?

?

Lost Person Behavior

Unit 5 – Using the Book

Slide 1

Using the book

Text Reference: ISRID Tables Explained page 75-90

Slide 2

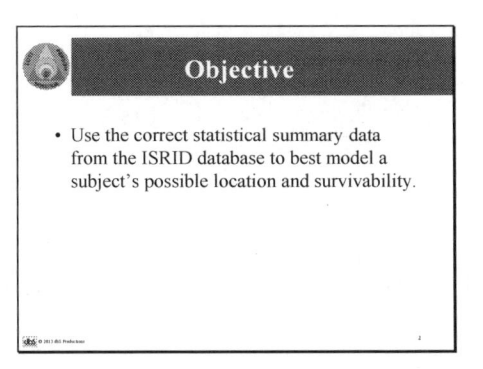

Objective
- Use the correct statistical summary data from the ISRID database to best model a subject's possible location and survivability.

While teaching this class many times, many students stated their goal was to better learn how to use the book on an incident. The goal of this unit is to show student how to do that and how to better interpret the tables.

Slide 3

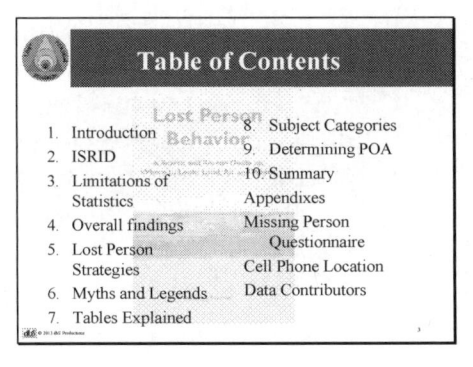

Table of Contents

Lost Person Behavior

1. Introduction
2. ISRID
3. Limitations of Statistics
4. Overall findings
5. Lost Person Strategies
6. Myths and Legends
7. Tables Explained
8. Subject Categories
9. Determining POA
10. Summary

Appendixes
Missing Person Questionnaire
Cell Phone Location
Data Contributors

Contents of the book

Lost Person Behavior — 85 — September 2014

Unit 5 – Using the Book

Slide 4

8. Subject Categories
- Direct Reference during a search
- 41 Subject Categories
- Organization
  - Subject Profile
  - Subject Statistics
  - Reflex Tasking
  - Additional Investigative Questions

The real purpose of the book is found in Chapter 8, which contains the subject categories. The chapter was intended to serve as a standalone reference. Where it would be possible to have all the information you might need on the subject in one place. Book was designed with binding to lay flat, so it would be easier to use in the field.

Slide 5

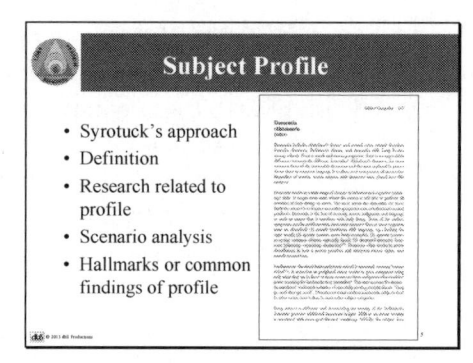

Subject Profile
- Syrotuck's approach
- Definition
- Research related to profile
- Scenario analysis
- Hallmarks or common findings of profile

What is contained in the subject profile.

Slide 6

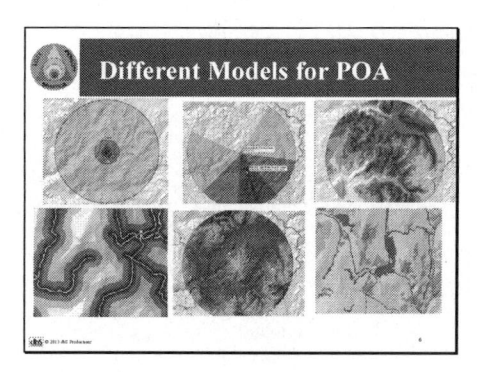

Different Models for POA

The slide shows the six statistical models that can be used help determine POA. From left to right (then top to bottom), the ring model, dispersion model, elevation model, track offset, mobility (along with travel cost), and feature model. All of these models will be discussed in more detail later.

Lost Person Behavior

Unit 5 – Using the Book

Slide 7

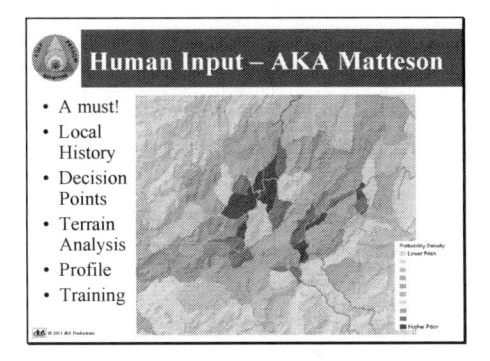

This slide is a repeat from the Myths and Legends unit (4). By default it is hidden. However, if this section is being presented as a standalone it should be displayed and discussed.

Slide 8

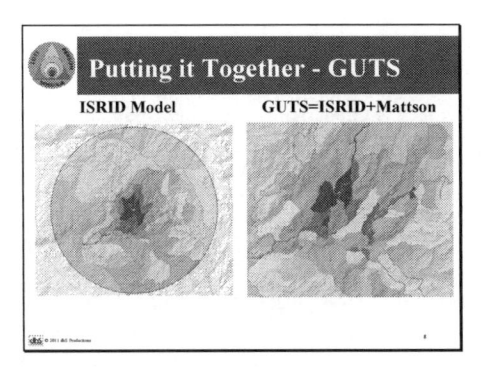

This slide is a repeat from the Myths and Legends unit (4). By default it is hidden. However, if this section is being presented as a standalone it should be displayed and discussed.

Slide 9

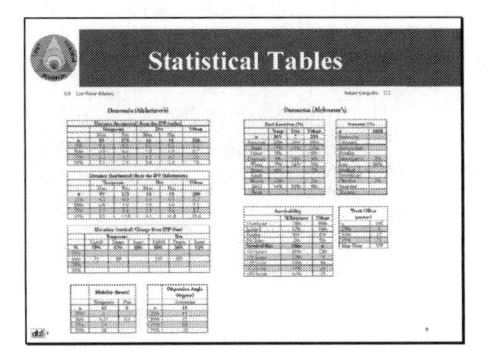

Example of the statistical tables. Each will be discussed in more detail. Instructor should focus on one of the subject categories when instructing about the tables. It does not have to be limited to Dementia.

Unit 5 – Using the Book

Slide 10

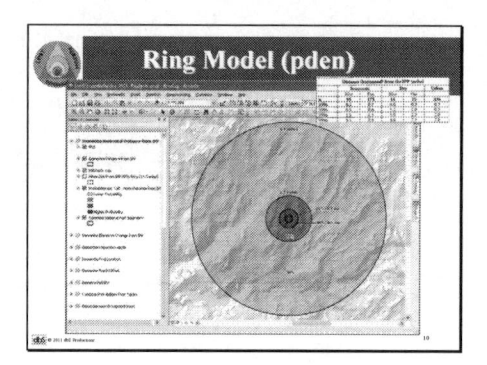

The concentric ring model has been used in search planning for years. The ring statistics are based on distance as the crow flies from the initial planning point (IPP) to the find location. The 50% zone is the median in which half of the cases were found. The 95% zone can be used to establish an early containment area for search planning.

For the Dementia category, 25% of the cases are found within 0.2 miles of the IPP, 50% of the cases are found within 0.5 miles of the IPP, etc.

For all of the tables, the probability calculation is done using raster data, not vector data. The result is probability per cell, i.e. there is x probability that a lost person will be found in this 10x10 cell. The cell size used in the analysis is 10 meters.

This statistics state that 50% of the cases were found 0.5 miles or less from the IPP (for the temperate ecosystem in mountainous terrain). In raster analysis the rings do not overlap, as is possible in vector analysis. Therefore, for all of the analyses presented here, the percentages have been changed to non-overlapping values, e.g. 25% of the cases are found in the 25% ring, 25% of the cases are found in the 25-50% ring, 25% of the cases are found in the 50-75% ring, and 20% of the cases are found in the 75-95% ring.

Lost Person Behavior

Unit 5 – Using the Book

Slide 11

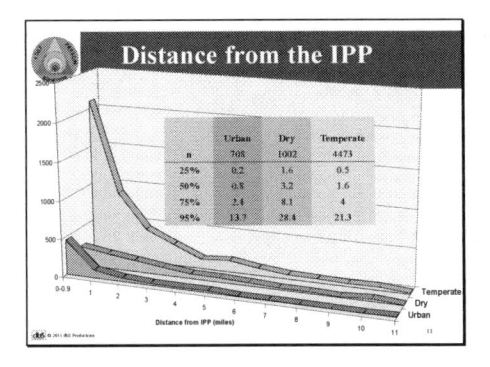

Slide 12

In an Urban and Dry ecoregion, the distinction between mountainous and non-mountainous does not seem to matter as much. In the Temperate ecoregion the classic Syrotuck finding of greater distance in mountainous terrain is supported. Currently statistical work is underway to formally evaluate these differences. Future editions of the book may not show terrain type for the Dry domain.

Slide 13

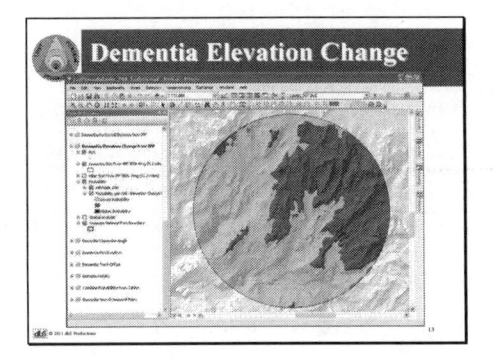

Probability per cell based on elevation change from the IPP to the find location for the Dementia category. The dark red shows areas with a higher probability for finding the subject, with the highest probability being at the same elevation as the IPP.

Unit 5 – Using the Book

Slide 14

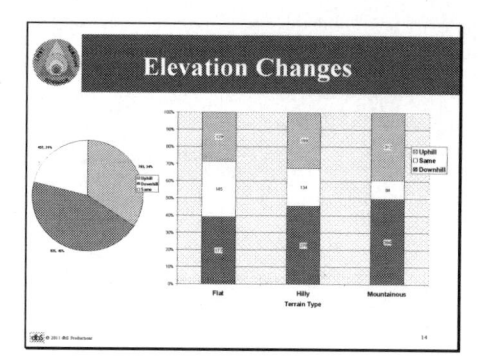

The more mountainous the terrain, the more likely the subject when up or down.

Slide 15

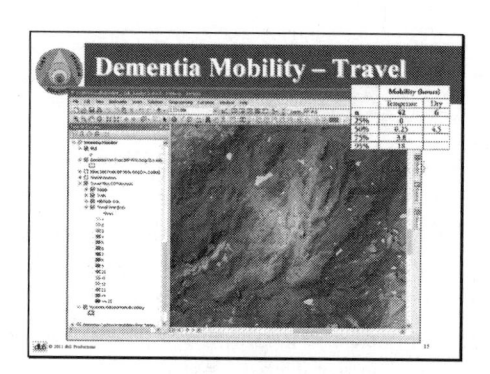

The next table is the Mobility table which show the amount of time the subject was moving. But before we can calculate the spatial extent of mobility, we need to create a travel time surface which shows us how far the person could go in x amount of time. Using a cross country mobility model that takes into account the difficulty to traverse various vegetation types, slope, water and roads and trails, we can create this travel time surface.

Each band or color represents an hour of walking. <u>Notice the lobes of speed</u>. You can see that the lobes of speed follow roads and trails where the subject could move more quickly.

Travel speed model is based on a model published on ArcScripts on April 22, 2008. I modified the model but used it principles. Original model:
Authors: Brent Frakes (brent_frakes@nps.gov), Courtney Hurst, David Pillmore, Billy Schweiger, Colin Talbert
Abstract: This document describes using the cost surface model in ArcGIS 9.2. This model uses a digital elevation model, streams, lakes, roads, trails and land cover to determine the relative time required to reach any location within a park unit. The

Unit 5 – Using the Book

result can be used to help stratify random sampling points to ensure that most occur in accessible areas.
Suggested Citation: NPS-ROMN (2008). Travel Time Cost Surface Model, Version 1.7, Rocky Mountain Network-National Park Service, Fort Collins, Colorado.
ArcScripts URL: http://arcscripts.esri.com/details.asp?dbid=15547

Slide 16

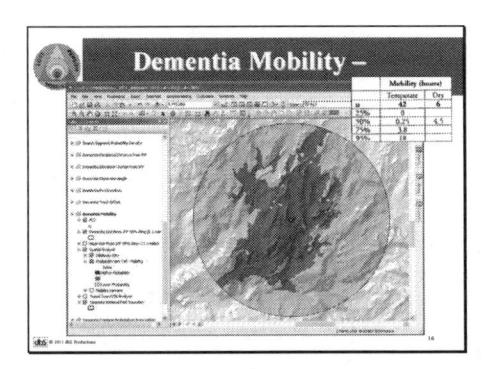

The mobility probability per cell result shows the higher probability areas in dark red.

Slide 17

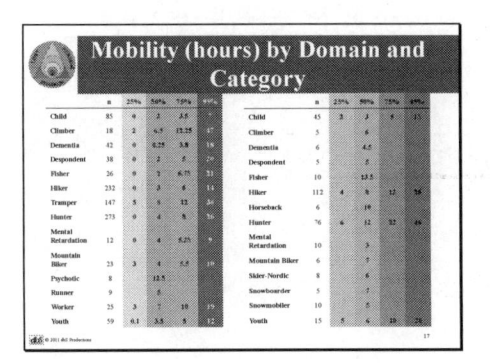

Table allows direct comparison between different profiles for temperate and dry domain.

Lost Person Behavior · 91 · September 2014

Unit 5 – Using the Book

Slide 18

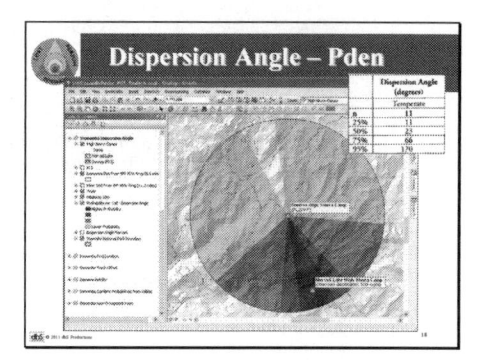

Probability per cell for the Dementia dispersion angle statistics. Darker red indicates a higher probability.

The 95% classification has been changed to 170 degrees based upon emails from Robert Koester.

25% of the subjects were found within 22 degrees (+/- 11 degrees from the direction of travel).

Note: these are not overlapping bands so you cannot say 95% of the hikers were found in the lightest red band, you can say that 20% of the lost hikers were found in the lightest red band and 25% were found in the darkest red band.

Slide 19

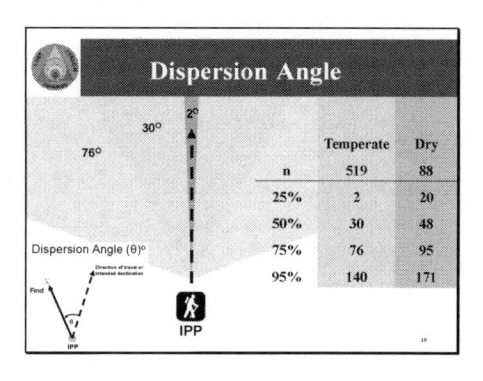

The intended destination is most useful for hikers. In the case of a circuit hike, then the destination was the point furthest away. For most of the cognitively impaired categories, a physical clue was required to establish the direction of travel. While track/trailing dogs are highly valuable, their trails were not used for the purpose of these statistics.

Lost Person Behavior

Unit 5 – Using the Book

Slide 20

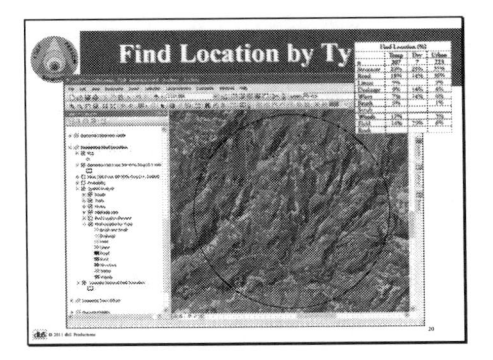

The find location is the actual location where the subject was found. This slide shows vegetation data combined with roads, trails, hydrology and buildings. The data has been reclassified into the Find Location categories. It is the data dataset used in the Hiker Find Location analysis.

In the temperate domain, only 9% of the dementia subjects were found along linear features which are trails, power lines, railroads, etc., as compared to 25% for the Hiker category. The Linear category does not include roads and drainages because they are their own category.

Slide 21

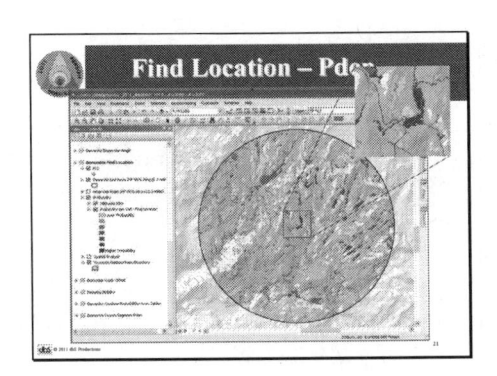

The probability per cell has been calculated. In the zoomed in area you can see a high probability for a meadow (Field type, 14%) next to the point last seen and a lower probability for trails (Linear type, 9%) and streams (Drainage type, 9%). The Woods classification had a high probability (17%) but because it covers a large area (most of the search area), it has a lower probability per cell. Also, rocky and barren areas in the landscape are not represented in the probability results because none of the ISRID dementia cases were found in that landscape type.

Unit 5 – Using the Book

Slide 22

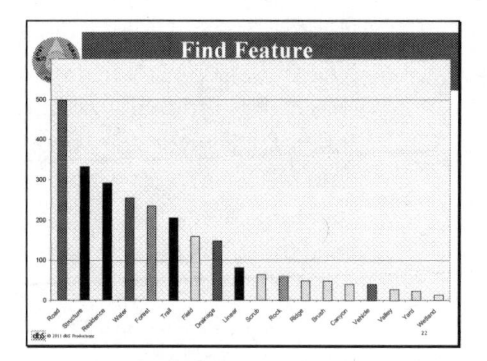

The actual database, contains 20 unique descriptions. However that was too many when it came time to make the tables. So the list was further combined to create the 10 seen in the table. For example vehicle, structure, residence, and yard were combined. Ice from mountains and glaciers were combined with rock. Currently, the book causes some confusion with linear, roads, and drainages. While roads and drainages are linear features they were further broken out.

Slide 23

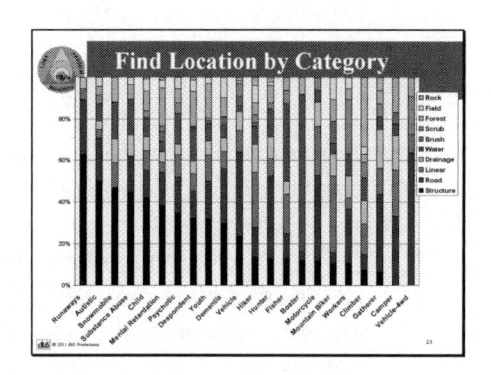

Different subject categories are more or less likely to be found in different areas.

Slide 24

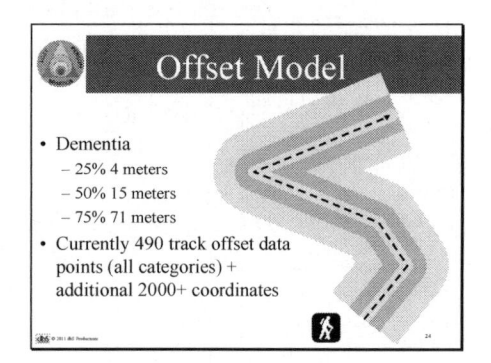

Track Offset is the perpendicular distance from the subject to the nearest linear feature. It may not even be the feature that subject used to get to the location.

Lost Person Behavior — September 2014

Unit 5 – Using the Book

Slide 25

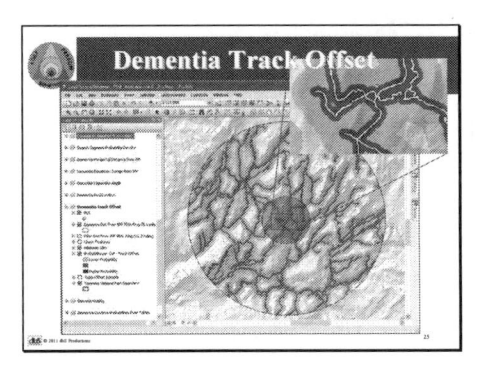

The probability per cell has been calculated for the track offset. Notice that the source linear features have been excluded from the calculation because they are already accounted for in the Find Location analysis.

In the Dementia category, the track offset values are small and this allows for corridor searching which can significantly decrease the area for searching, especially in the large 75%-95% zone.

Slide 26

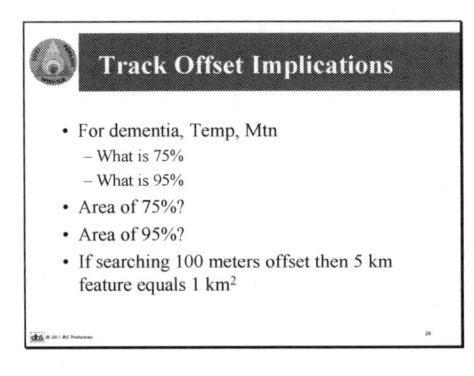

Purpose of the math is two-fold. First get the class using the book to look up the actual numbers. It also illustrates the value of cooridor tasks taking place in the 75-95% zone. To simplify the math, use miles, round 1.2 miles to 1 miles (for 75%). Round 5.1 to 5 (for 95%). Simplify pi to 3. Then the area of the 75% becomes 3 sq Miles, the area of 95% becomes 75 – 3 = 72 sq Miles. An area of this size cannot be grid searched. However, a corridor search of a road 100 meters on either side (200 meters) that was 5 km long would only require searching a 0.2 x 5 = 1 sq km. This can be achieved. Navigation and safety concerns are not as high when conducting corridor searches.

Lost Person Behavior   September 2014

Unit 5 – Using the Book

Slide 27

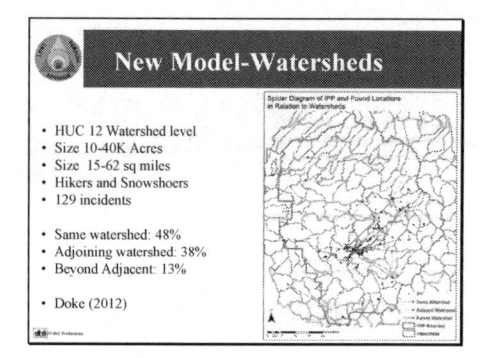

New Model-Watersheds
- HUC 12 Watershed level
- Size 10-40K Acres
- Size 15-62 sq miles
- Hikers and Snowshoers
- 129 incidents

- Same watershed: 48%
- Adjoining watershed: 38%
- Beyond Adjacent: 13%

- Doke (2012)

Watershed typically would need to be downloaded into GIS software. However, it represents an approach that better matches features that may actually constrain a subject. The size of a HUC-12 watershed is similar to a SAR planning region (15-62) Sq miles. A HUC-12 watershed can also easily be cut in half by using the actual water feature as a boundary for a smaller planning region.

Slide 28

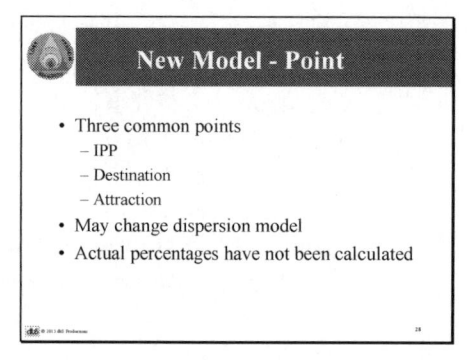

New Model - Point
- Three common points
  - IPP
  - Destination
  - Attraction
- May change dispersion model
- Actual percentages have not been calculated

The update of ISRID is working on new models

Slide 29

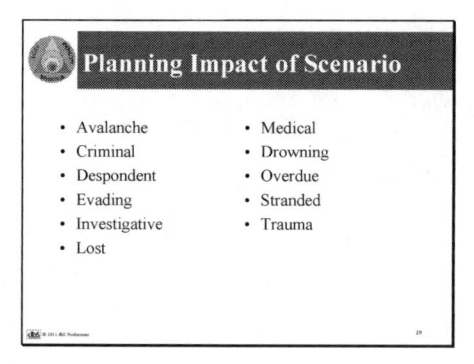

Planning Impact of Scenario
- Avalanche
- Criminal
- Despondent
- Evading
- Investigative
- Lost
- Medical
- Drowning
- Overdue
- Stranded
- Trauma

The formal definitions are given on page 86 of the book. Avalanche is self-explanatory. Criminal and Despondent are unique in each being a subject category. Evading requires different tactics/resources and increases the need search an area multiple times. Investigative is a non-search resource. Lost is the focus of the course for the most part. Medical requires special mention. The subject most likely is not lost and on the intended route. However, most ill subject feel vulnerable and will not sit down right on the route. Instead most will move a short distance off the route and sit against something so they are more comfortable and not feel vulnerable from behind. Some medical conditions

Lost Person Behavior   September 2014

Unit 5 – Using the Book

(hypoglycemia, strokes, etc) can lead to irrational behavior. Drowning is self-explanatory and is covered by the water category. Overdue implies they were not lost and will be found on the route. However, being lost and then becoming reoriented with a time delay is often reported as overdue. Stranded means the person was trapped and unable to move due to an external force. Often means subject on intended route or moved to higher ground to avoid flood waters. Trauma also suggests the subject will be close to the route of travel. However, most subject much like medical will move a short distance if able.

Slide 30

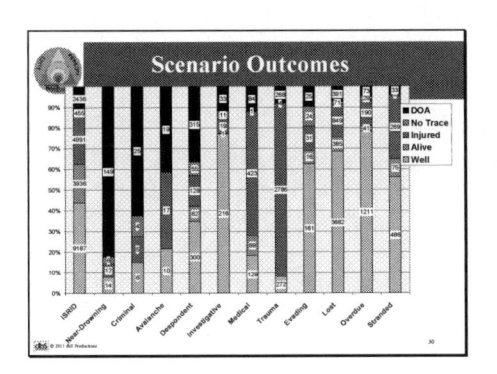

Different scenarios have different outcome probabilities.

Unit 5 – Using the Book

Slide 31

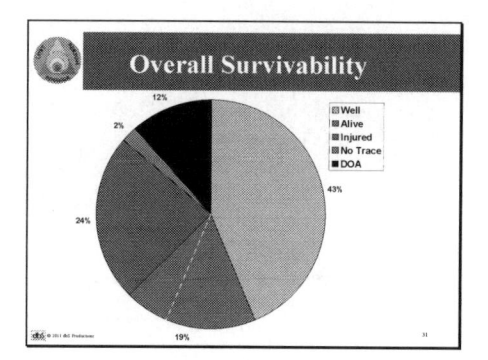

Overall Survivability provides little useful information other than to give an overall picture. Most data contributors classified the subject as well, injured, or DOA. However, some just used Alive or DOA. Therefore the alive contains both injured and well. Most likely in roughly the same proportion. This would make injured 31% and well 55%

Slide 32

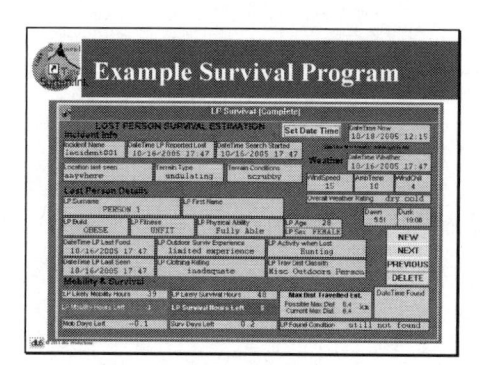

The following is given as a sample of the only known land specific survivability predicting tool on the market (several maritime models exist). The model gives an absolute number of hours the subject is expected to survive. The model is based upon Kelly's 1973 data. A better approach would be to give probabilities of survival. The program is not available and comes out of Australia

Slide 33

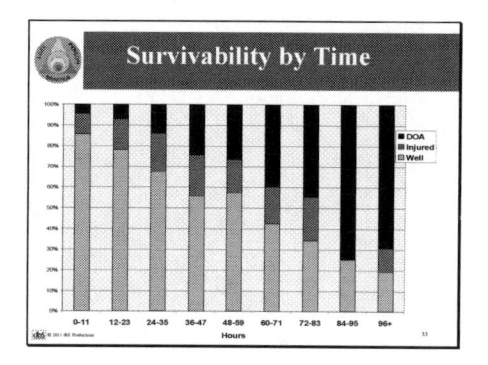

Survivabiliy was discussed in the Myths section. It is repeated here to show the importance of time. Each bin is based upon the time the subject was last seen and when they were found. Each time bin was looked at separately. Each bin as time progresses as less and less data. The 84-95 illustrates that with less and less cases it appears the number of injuries goes away. Yet looking at the everything more than 96 hour bins which includes more cases we get a more expected result.

Lost Person Behavior   September 2014

Unit 5 – Using the Book

Slide 34

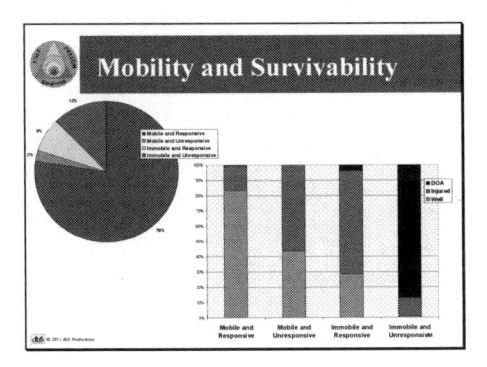

Mobility and Survivability

Slide 35

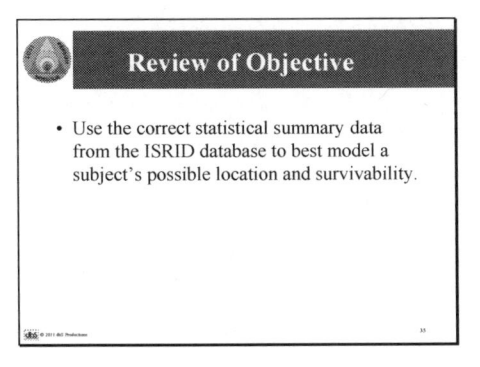

Review of Objective

- Use the correct statistical summary data from the ISRID database to best model a subject's possible location and survivability.

Slide 36

Questions?

?

Lost Person Behavior — 99 — September 2014

Unit 6 – Reflex Tasking

Slide 1

Reflex Tasking – The Bike

Text Reference: Unit 9 Page 295-301. Search Wheel can also be used as a reference. In this unit we will discuss and demonstrate the following; problems with the classic approach to initial tasks, who reflex tasking can be used, what information is required for reflex tasking, the use of the bike model, the use of the search wheel job aid, and finally a demonstration of the techniques.

Slide 2

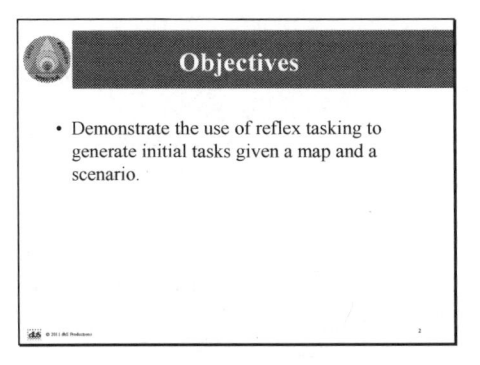

Objectives
- Demonstrate the use of reflex tasking to generate initial tasks given a map and a scenario.

Participants will have the opportunity to practice using the bike model and reflex tasking in two map problems at the end of this unit. The first may be done as a group, the second individually.

Slide 3

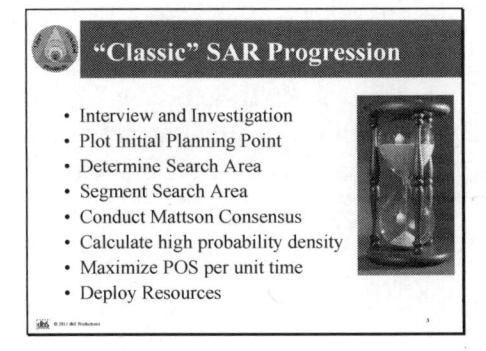

"Classic" SAR Progression
- Interview and Investigation
- Plot Initial Planning Point
- Determine Search Area
- Segment Search Area
- Conduct Mattson Consensus
- Calculate high probability density
- Maximize POS per unit time
- Deploy Resources

In most classic SAR management textbooks a standard process in taught to new students. It is typically taught as a step-by-step process. In reality most experienced SAR managers don't follow the steps. However, new students often follow the process and it may take 1-4 hours to get through all the steps. In one case 200 resources waited in staging for four hours while management completed all the steps. This wasted 800 man-hours of potential search effort. As an instructor it became clear that the method that better reflected the actual process used by more experienced SAR managers be followed and taught. The classic SAR progression still have a place and great value when sufficient time is available for formal

Lost Person Behavior   September 2014

Unit 6 – Reflex Tasking

planning or when initial tasks are already complete.

Slide 4

Purpose of Reflex Tasking

- Rapid planning and deployment process
- Reflects process of experienced planners
- Sends resources to higher probability areas
- Tailored for subject category and situation
- Reflex ≠ Hasty
  - In some cases may include grid/sweep tasks
  - In some cases may include corridor tasks

Reflex tasking was developed to better reflect the actual process used to deploy resources by experienced SAR planners and operations personnel. It recognizes the tactics used at the beginning of an incident are often the same or falls into general patterns. One of the major purposes of reflex tasking is it allows rapid planning. It is possible to generate a dozen tasks within 10 minutes with experience.

Slide 5

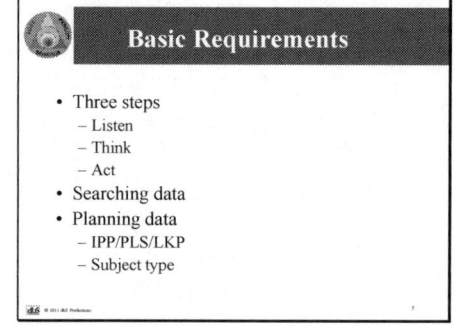

Basic Requirements

- Three steps
  - Listen
  - Think
  - Act
- Searching data
- Planning data
  - IPP/PLS/LKP
  - Subject type

Many different planning processes have been described and all are correct. The process listed here is for reflex tasking. This is the minimal information you must have to start the planning process.

Unit 6 – Reflex Tasking

Slide 6

Reflex Planning Process
- Ten Steps following the Bike Model
- Use Reflex Tasking Worksheet or Bike Wheel Model on page 299 of LPB book
- Brainstorm process
- Just do, get task ideas down on map
- Try to create more tasks than resources

Instructor should pass out handout or refer student to student workbook. It is highly encouraged to use the worksheet the first time the process is used. The class will explain the process first, give an example, and then allow the student to practice with two scenarios.

Slide 7

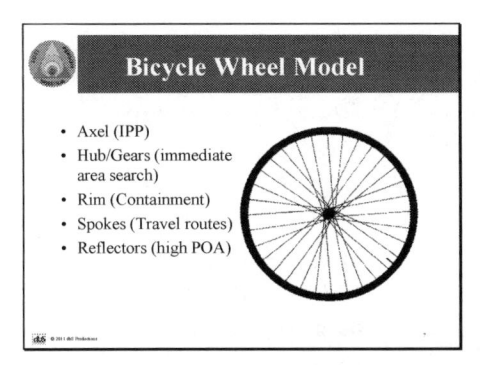

Bicycle Wheel Model
- Axel (IPP)
- Hub/Gears (immediate area search)
- Rim (Containment)
- Spokes (Travel routes)
- Reflectors (high POA)

A model that can help a student remember the reflex task process in the bike wheel model. Think of the back wheel of a geared bike.

Slide 8

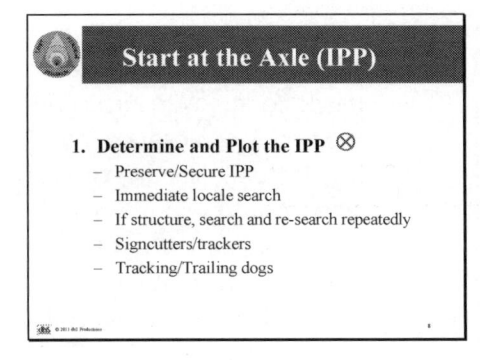

Start at the Axle (IPP)

1. Determine and Plot the IPP ⊗
   - Preserve/Secure IPP
   - Immediate locale search
   - If structure, search and re-search repeatedly
   - Signcutters/trackers
   - Tracking/Trailing dogs

The next eight slides describe the ten-step process to create reflex tasks. At the IPP itself, six different tactical tasks can be deployed.

Lost Person Behavior

Unit 6 – Reflex Tasking

Slide 9

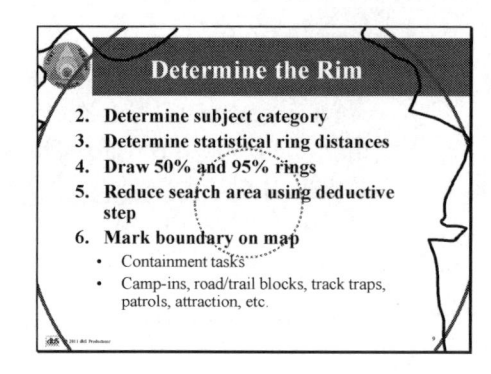

Determine the Rim
2. Determine subject category
3. Determine statistical ring distances
4. Draw 50% and 95% rings
5. Reduce search area using deductive step
6. Mark boundary on map
   - Containment tasks
   - Camp-ins, road/trail blocks, track traps, patrols, attraction, etc.

The next steps are all related to determining the out search boundary. It may be argued that the hub could come before the rim, but nevertheless both need to be determined soon. For many searches it is containment that might be the most critical component. Ken Hill has shown that the subject that travel the furthest, often are walk outs.

Slide 10

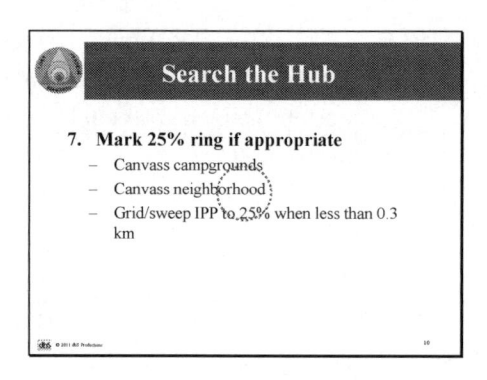

Search the Hub
7. Mark 25% ring if appropriate
   - Canvass campgrounds
   - Canvass neighborhood
   - Grid/sweep IPP to 25% when less than 0.3 km

The hub is often defined by the 25% ring. In many cases it is 300 meters (0.3 km) or 328 yards. For some subject types (dementia, children, despondent) a sweep or grid task within the 25% may be appropriate almost right away. For other subject categories it may not make sense. The number of some subject categories such as hiker may be more a reflection of hikers who return to the destination.

Slide 11

Hub Distances

Category	Km	POA%
Children	0.3	50%
Despondents	0.3	50%
Dementia	0.3	25%
Intellectual disability	0.3	25%
Mental Illness	0.6	25%
Hikers	0.6	25%
Hunters	0.6	25%
Autism	0.6	25%

Table showing the 25% or 50% of some subject categories. Children and despondents warrant a grid search within the first 300 meters early. Dementia and Intellectual disability (mental retardation) also warrant an early sweep/grid task within the first 300 meters. Keep in mind, nothing is absolute about 300 meters. If terrain features suggest searching to 350 meters or only to 250 meters, then that should make sense.

Lost Person Behavior — September 2014

Unit 6 – Reflex Tasking

Slide 12

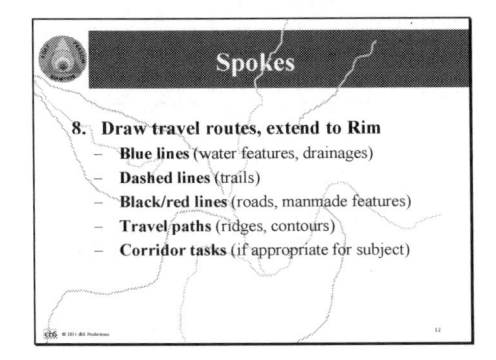

Searching the spokes may be one of the most effective search techniques. Spokes include searching blue lines or water features or land features where water might flow given sufficient rain. All manmade features such as trails, roads, power lines, railroad tracks, pipeline cuts, etc should be tasked. In addition possible travel paths such as ridgelines (common for some hunters) or contours that that follow the start of steep terrain should be considered. For the hiker category 75% (dry) or 58% (temperate) of hikers are found on a spoke feature. Tactically it can be carried out my a large variety of field resources. All of which can be small mobile teams.

Slide 13

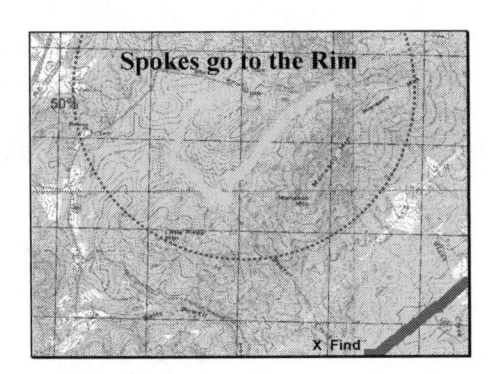

REFLEX TASK to SEARCH BOUNDARY. At least draw up those tasks even if lacking resources. This is the reason the rim (search boundary) is determined prior to drawing the spokes. Search for a missing 10-year old female. Last seen at her grandmothers house taking a nap. Went out to play with family dogs. Temperature during the day 70 F, at night fell to freezing temperatures. Initial reflex tasking during the evening took advantage of C shape of the stream. One of the family dogs returned the following morning. Beaver Creek was not searched until the second full day. She was located deceased.

Lost Person Behavior   September 2014

Unit 6 – Reflex Tasking

Slide 14

Reflector

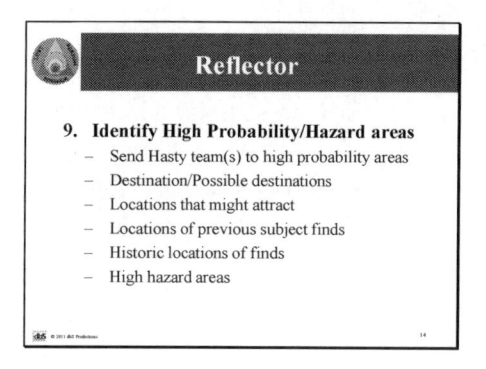

9. Identify High Probability/Hazard areas
   - Send Hasty team(s) to high probability areas
   - Destination/Possible destinations
   - Locations that might attract
   - Locations of previous subject finds
   - Historic locations of finds
   - High hazard areas

The reflector or perhaps better called attractor is any area in the search area that has higher probability. Think candy store, playground, or water for children. Former residence or workplace for dementia, tree stand or favorite hunting spot for a hunter, scenic view for despondent. It also includes areas of high hazards. This can be important for children, dementia, autistic, and similar categories which may not perceive danger correctly.

Slide 15

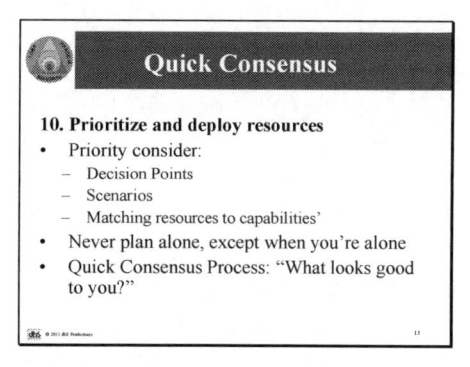

Quick Consensus

10. Prioritize and deploy resources
    - Priority consider:
      - Decision Points
      - Scenarios
      - Matching resources to capabilities'
    - Never plan alone, except when you're alone
    - Quick Consensus Process: "What looks good to you?"

The quick consensus process is an informal non-mathematical process. It can be accomplished many different ways. Tasks could already be drawn up and the second person determines which tasks should be sent out first. It could be as simple as a team leader selects and comments on a task. It can also be done, by asking which areas look good prior to writing up any of the reflex tasks.

Slide 16

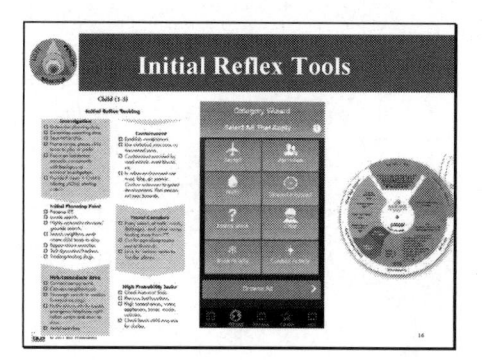

Initial Reflex Tools

The following are tools to help with the reflex task process. Within the book (Lost Person Behavior) is a list of reflex tasks for each subject type. A Lost Person Behavior APP (currently in beta development) also provides the same information (battery not included). Finally, the search wheel also describes the process and provides the statistics for the seven most common subject types. New search wheels will become available for Temperate, Dry, and Urban domains.

Lost Person Behavior

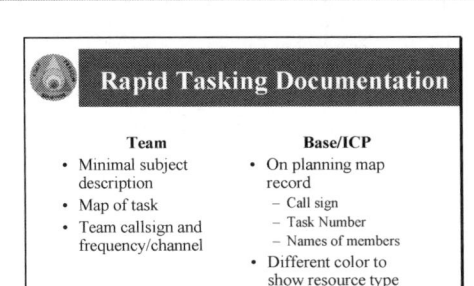

Unit 6 – Reflex Tasking

Slide 17

Rapid Tasking Documentation

Team
- Minimal subject description
- Map of task
- Team callsign and frequency/channel

Base/ICP
- On planning map record
  - Call sign
  - Task Number
  - Names of members
- Different color to show resource type assigned.

While this course does not fully address search management, nonetheless documentation and information processing is critical. At a minimum the search team must have a description of its task, a map, and communication details, and a description of the subject. Even if it is as simple as bring back every two-year old you find in the middle of the woods alone at night. The command post must know where it has sent every team, who is a on every team, and how to communicate with every team.

Slide 18

Matthews Arm Case Study

- Fit 82 year old male, last seen at Matthews Arm Campground at 10:00. Reported missing at 19:30. Current time 20:00
- Normally walks 3 miles everyday in Sweden
- Mild dementia (Alzheimer's disease)
- Was missing for a few hours yesterday.

Scenario: Here you will be responsible for searching for a missing mild dementia subject out hiking. You have been called to Matthews Arm Campground for a fit missing 82 year old male last seen today at 1000. He normally walks 3 miles (5 km) everyday in Sweden. He is visiting Shenandoah National Park along with his family. He was also missing for a few hours yesterday but wandered back into the camp. The point last seen is indicated on the map below. He was reported missing by family members at 1930 when the dog he went out to walk returned to the campground. The current time is 20:00 It is a warm August day in the park. Highs in the 80's and nighttime lows in the 50's. It is currently clear and warm.

Lost Person Behavior

Unit 6 – Reflex Tasking

Slide 19

Point out red dot represents the IPP. Ask class "What do we desperately need to do now!" Reply should be **limit the search area**.

Slide 20

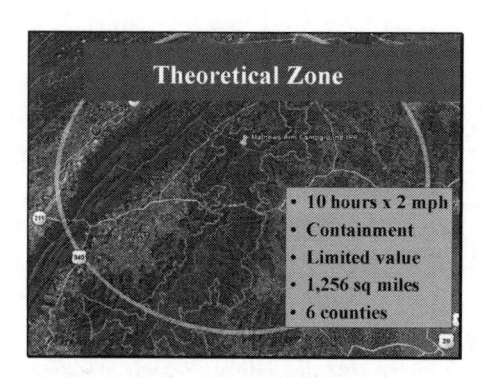

It can be quickly seen that the theoretical zone as little value in limiting the search value if the search is delayed more than a few hours. In purple is the statistical zone as well.

Slide 21

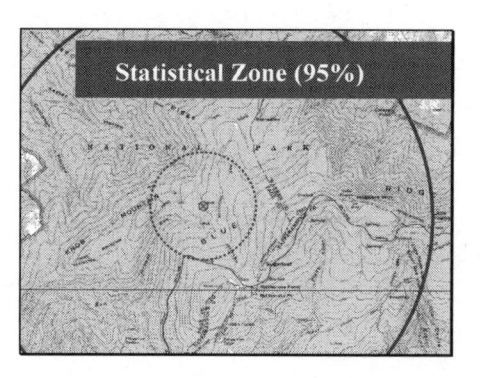

95% zone or Max zone and 50% zone are drawn on the map.

Lost Person Behavior

Unit 6 – Reflex Tasking

Slide 22

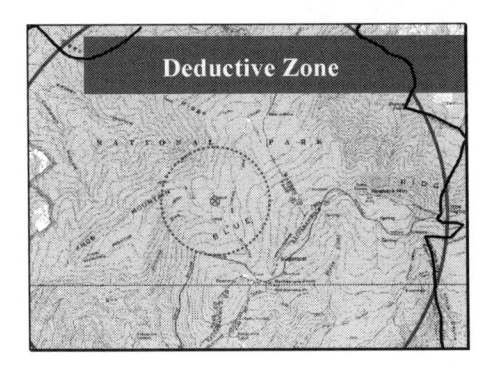

The final search boundary is determined not by the statistical circles but by land features that a team will be able to locate in the field. This step may also cause the search area to shrink for areas the subject is not likely to cross, or grow slightly if no land feature can be found near the circle. This is covered in a basic search management course.

Slide 23

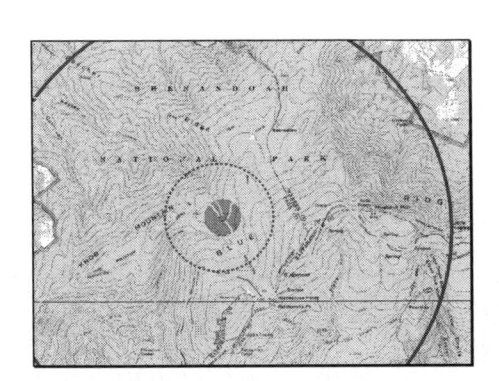

What really happened: The search occurred in the early 80's before Alzheimer's or dementia had become a common buzz word. The subject typically walked 3-4 miles (5-7 km) in his native land of Sweden. His family reported they noticed he seemed much more confused during the last 8 months. The family was visiting the United States and as part of the visit went camping at Shenandoah National Park in Virginia (please note: it is not recommended to take Alzheimer's disease patients camping in the wilderness!). Initial hasty tasks focused on searching the camp ground and running trails. An air-scent dog had a weak alert during an initial night task at the intersection of the jeep trail and Keyser Hollow run drainage. A follow up hasty task down the drainage then made the find first thing the following morning. The subject was in good condition. He stated he simply had gotten confused. He was able to walk out of the woods unassisted. It is not clear how he had gotten to his location.

Unit 6 – Reflex Tasking

Slide 24

Still no Subject?

- Repeat tasks initially performed at night once daylight
- Repeat tasks if high likelihood of mobile subject
- Continue area tasks to slightly beyond 50%
- Consider corridor tasks if appropriate
- Formal search segmentation and consensus.

Reflex tasking will not always find every subject. The slide provides a few more suggestions prior to switching over to formal search theory which requires search segmentation and a formal consensus process.

Slide 25

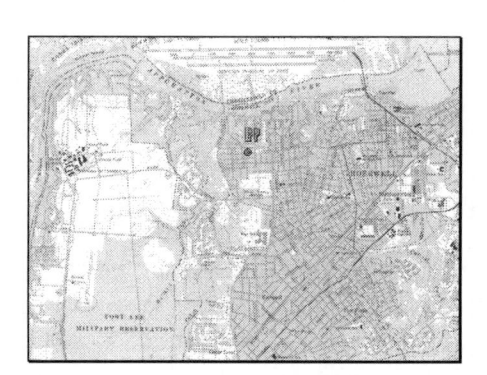

68 year old male suffers from moderate Alzheimer's disease. He has been the subject of three other searches. In all three searches he departed from home and was found at the location indicated on the map. The subject was reported missing by his wife at 17:30. He has been missing for 30 minutes. He is in average physical condition for his age. It is February, highs in the 50's, lows in the upper 30's.

Slide 26

Theoretical Search Area
Radius = Time (hrs) x Speed (mph)

Lost Person Behavior   110   September 2014

Unit 6 – Reflex Tasking

Slide 27

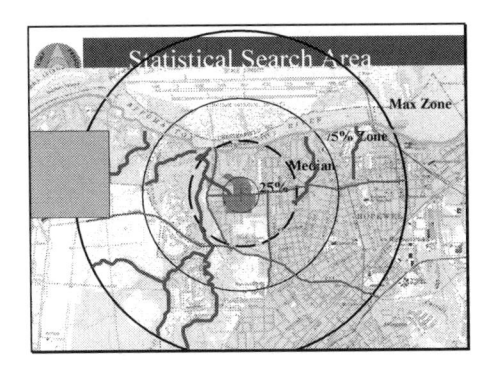

Clues:

1. Locations where he was previously located.
2. Sighting: A neighbor sighted him walking west along the road he lives on. High reliability of sighting based upon clothing and time. Information requires neighborhood canvas or media release. (37.3017993 N, 77.3213917 W)
3. Trailing dog: A scent-discriminating German Sheppard follows the track indicated on the map. The team has a reputation for both outrageous claims and finding subjects in previous searches.
4. Air-scent dog alert: Air-scent dog team alerts. Wind is non-directional and seems to be following local micro conditions. (37.3017303 N, 77.3241838 W)
5. **Find location** (37.3017647 N, 77.3241409 W)

What actually happened?

Search resources were contacted at night. The trailing dog did report the track presented in the map problem. However, this search occurred before the existing training standards were developed. An air-scent dog team did have the dog alert in cabin creek. Since the alert occurred at night it was followed up with a second dog task the following morning. It is a good practice to repeat all reflex tasks that occur at night again in the light. The subject was found alive by this team early in the morning. The subject was buried in mud up to his chest. He had a core temperature of 85o and was airlifted to the hospital. He recovered, as neurology intact as when he first became missing.

Learning Points: Need to canvass neighborhood in urban searches. Value of reflex tasking. Follow up on night time tasks. Careful evaluation of all tracking clues.

Unit 6 – Reflex Tasking

Slide 28

Reflex Tasking Assignment
- For each scenario use the bike wheel model to create tasks for the axle, hub, rim, spokes, and reflectors.
- Mark your tasks on the map
- You will be given 10 minutes for each map
- You may use the textbook and worksheet

See Instructor's Activity Guide for a full description

Slide 29

Scenario 1
- 2 19-year-old females (college roommates), last seen at old family friend's house (IPP)
- One lives locally the other is visiting.
- Traveling on foot. Came from home, told family friend headed back home when left.
- Missed a dinner date with parents. Described as highly responsible. Reported missing at 21:00. It is summer time.

It is 00:05 a warm summer night in June. Two 19 year old females were last seen at 17:30 leaving a friends residence (IPP). The two should have returned home (indicated by red circle) at 18:30 to go out to dinner with other friends. They are both students at Brigham Young University and are described as extremely responsible and reliable. The students are roommates at the university. One student has lived at the home location for 8 years. The second roommate is from out of state.

Slide 30

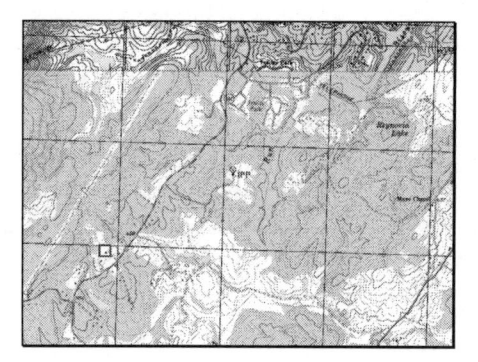

What really happened: The two friends decided to take a short cut in order to visit the horse stables at location A. They quickly became stuck in the thickets. Once night fell, they decided to stay put until morning. They were located by the first task dispatched using reflex tasking principles.

Lost Person Behavior

Unit 6 – Reflex Tasking

Slide 31

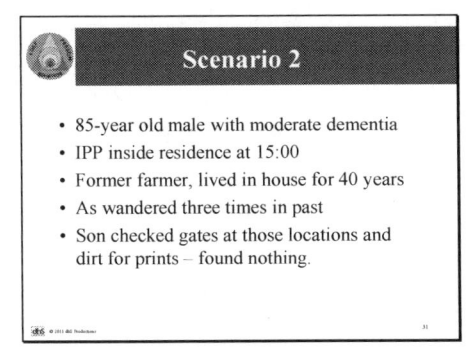

Scenario 2
- 85-year old male with moderate dementia
- IPP inside residence at 15:00
- Former farmer, lived in house for 40 years
- As wandered three times in past
- Son checked gates at those locations and dirt for prints – found nothing.

85 year old male with moderate dementia was last seen at his residence at 15:00. The former farmer has lived in this house for the past 40 years. He is taken care of by his son and daughter-in-law. He has been lost on three previous occasions. This time the son checked the gates that lead to the locations where he was previously found. He reports the gates where all closed and that he checked the dirt in front of each gate for footprints. He reports he did not see any footprints. It is June with a temperature of 75, no chance of precipitation, and night time lows of 65.

Slide 32

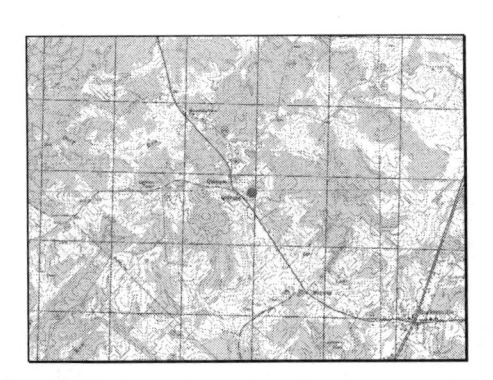

Fifty resources from the local rescue squad, fire department, and sheriff's department were quickly mobilized. Most of the resources were sent to locations where the man was previously found, essentially ignoring the information provided by the son. However, some resources were sent across route 33 in order to check the drainage. At the time, as indicated on the map, route 33 was a two lane highway. It currently is a four-lane divided highway. Nonetheless, always look on the opposite side of roads. Alzheimer's disease subjects have managed to walk across interstates. The subject was in fact located directly down his driveway across the road. He was found in good shape, but trapped in briars. The team was only 5 minutes into its tasks when it made the find. From the arrival of search management the search only lasted 23 minutes.

Unit 6 – Reflex Tasking

Slide 33

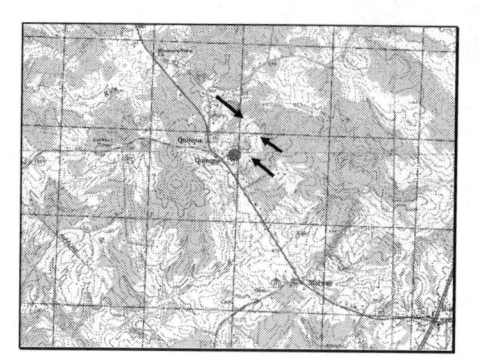

Fifty resources from the local rescue squad, fire department, and sheriff's department were quickly mobilized. Most of the resources were sent to locations where the man was previously found, essentially ignoring the information provided by the son. However, some resources were sent across route 33 in order to check the drainage. At the time, as indicated on the map, route 33 was a two lane highway. It currently is a four-lane divided highway. Nonetheless, always look on the opposite side of roads. Alzheimer's disease subjects have managed to walk across interstates. The subject was in fact located directly down his driveway across the road. He was found in good shape, but trapped in briars. The team was only 5 minutes into its tasks when it made the find. From the arrival of search management the search only lasted 23 minutes.

Slide 34

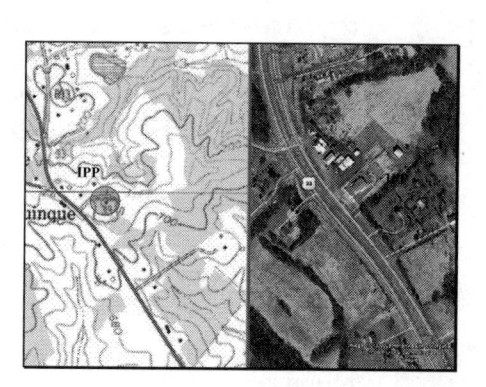

Close-up of IPP. Note the topographic map on the left shows the search area as it existed at the time of the search. The Google map shows route 33 as a four-lane divided highway, which was NOT the case at the time of the search.

Lost Person Behavior

Unit 6 – Reflex Tasking

Slide 35

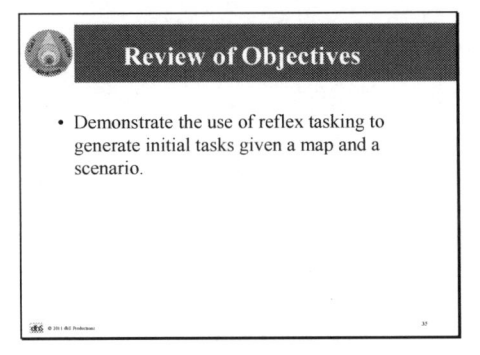

Review of Objectives
- Demonstrate the use of reflex tasking to generate initial tasks given a map and a scenario.

Slide 36

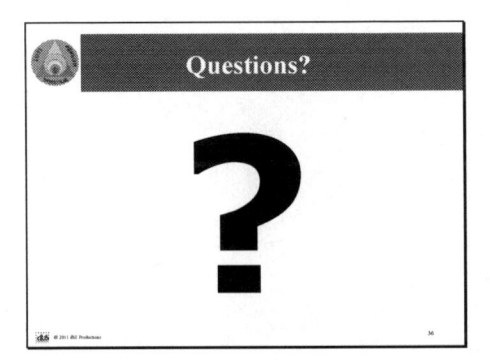

Questions?

Ask for questions

Lost Person Behavior — September 2014

Unit 6 – Reflex Tasking

Unit 7 – Subject Categories

Slide 1

Subject Categories

Text Reference: Chapter Eight. When instructing some subject categories may be skipped to best meet the needs of the class. The material provided in the slide show with all of the map problems will require more than 8 hours.

Slide 2

Objectives
- Describe key points from each subject category profile.
- Demonstrate the ability to deploy resources appropriate for each subject category given a map and scenario information.

Slide 3

Abductions
- Most data for children
- 60% start as search
- 4% urban/0.3% other
- 16% found alive
- Adults vs Child
- Concealing body
- PLS, Initial contact, assault site, murder site, body site
- Dumping site- quick, easy, vehicle access, secluded, downhill, natural cover
- Red Flags- white female, 5-12, missing from familiar location, no history of runaway, disappearance unexplained.

Text reference: Page 93-98 Abduction data comes from two sources. The primary source is the Brown et al report often called the Washington Study which examined 735 Children Stranger abduction homicides in the United States. A copy of this study is included in the Instructor resource DVD. The ISRID database also contains abduction data but only 15 cases with distance data. A key resource for students interested in additional information is material and classes offered by the National Center for Missing & Exploited Children (NCMEC). In this section four cases are presented that resulted in major searches. The cases should be presented and participants asked to draw

Lost Person Behavior — September 2014

Unit 7 – Subject Categories

conclusions.

Slide 4

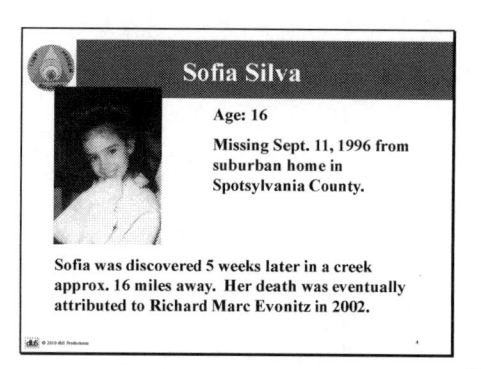

Sofia was last seen sitting on her front porch. Family was at home (inside) at the time of her disappearance. Limited search effort approx 1.5 mile radius around the home. Emphasis on investigation, but no solid leads.

Slide 5

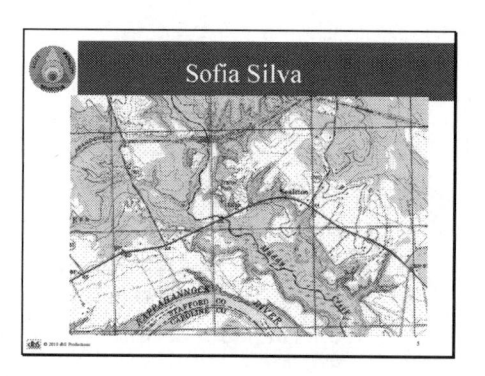

Her body was discovered while workers were clearing beaver dams which were causing water to flood the local business property. Notice ease of access, lack of population, and proximity to water.

Unit 7 – Subject Categories

Slide 6

Sofia Silva

Her body was discovered while workers were clearing beaver dams which were causing water to flood the local business property. Notice ease of access, lack of population, and proximity to water.

Slide 7

Sofia Silva

Her body was discovered while workers were clearing beaver dams which were causing water to flood the local business property. Notice ease of access, lack of population, and proximity to water.

Slide 8

Kristin & Katie Lisk

Ages: 15 & 12

Missing May 1st, 1997 from rural home in Spotsylvania County.

The Lisk sisters were found in the South Anna River 5 days later approx. 30 miles away.

Their deaths were also attributed to Richard Marc Evonitz in 2002.

At the time, this was the largest search in Virginia history. The formal search effort lasted 3 days, covered 75 sq miles (5 mi radius), and involved nearly 1200 personnel.

Lost Person Behavior

Slide 9 — Ease of access, proximity to water, and remote location.

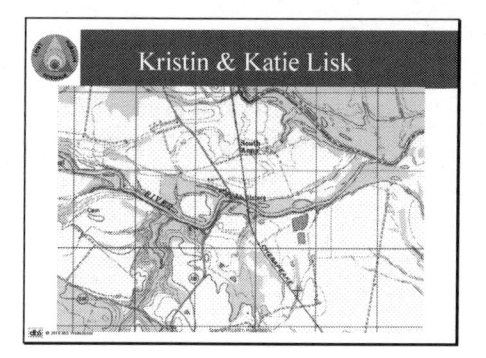

Slide 10 — Ease of access, proximity to water, and remote location. The picture was taken after the road was upgraded. At the time a more distinctive pull off existed on the right hand side of the road, that was often used by anglers.

Slide 11 — Ease of access, proximity to water, and remote location.

Unit 7 – Subject Categories

Slide 12

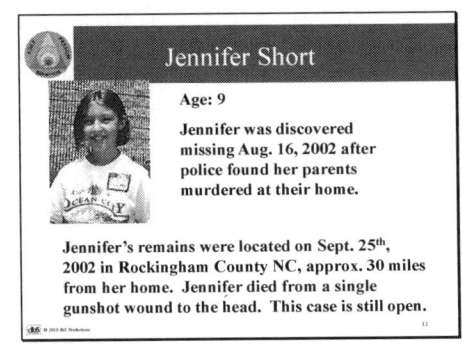

Garrison Bowman was detained by Canadian police on unrelated charges, while the FBI worked to obtain a material witness arrest warrant and arrange for his extradition back to Virginia. Bowman appeared before a closed grand jury session in Roanoke. No indictment was handed down. This case is still under investigation.
September 2003 – A ceremony was held to dedicate an un-named bridge near the find site in Jennifer's name.

Slide 13

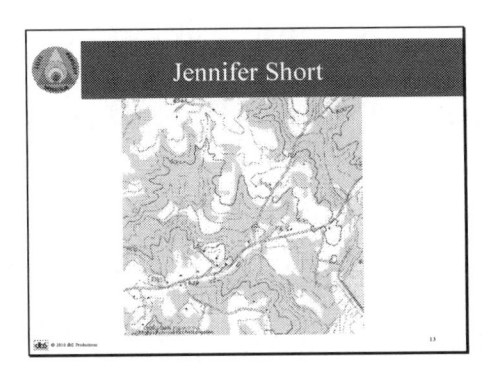

Jennifer's remains were discovered under a bridge on Grogan Rd by a man and his dog.

Slide 14

Jennifer's remains were discovered under a bridge on Grogan Rd by a man and his dog.

Lost Person Behavior   September 2014

Unit 7 – Subject Categories

Slide 15

Jennifer's remains were discovered under a bridge on Grogan Rd by a man and his dog.

Slide 16

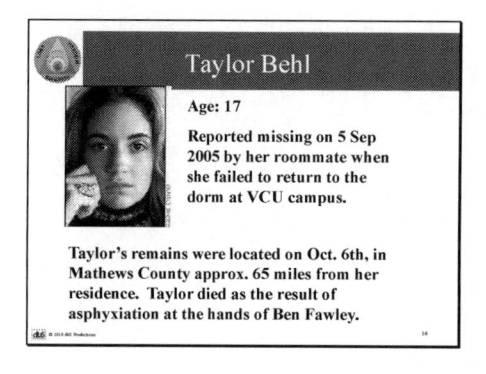

Slide 17

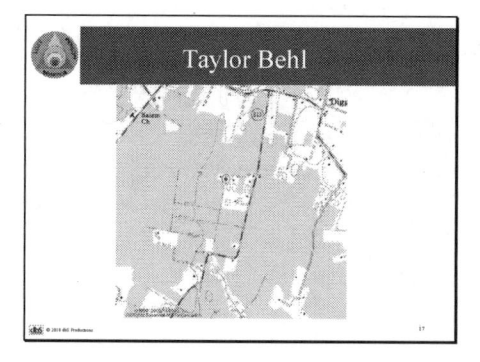

Lost Person Behavior — September 2014

Unit 7 – Subject Categories

Slide 18

Slide 19

Slide 20

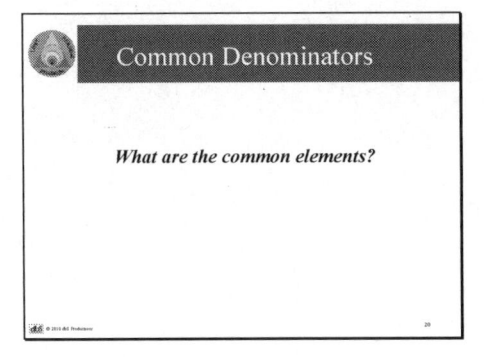

Numerous laws have been passed regarding lost and missing children. Like most good federal laws, they were inspired by tragedy at the local level. Law enforcement and SAR personnel should be keenly aware of three critical pieces of legislation.

Slide 21

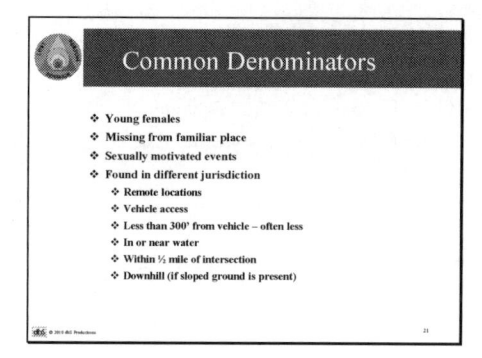

Common Denominators

- Young females
- Missing from familiar place
- Sexually motivated events
- Found in different jurisdiction
  - Remote locations
  - Vehicle access
  - Less than 300' from vehicle – often less
  - In or near water
  - Within ½ mile of intersection
  - Downhill (if sloped ground is present)

Numerous laws have been passed regarding lost and missing children. Like most good federal laws, they were inspired by tragedy at the local level. Law enforcement and SAR personnel should be keenly aware of three critical pieces of legislation.

Slide 22

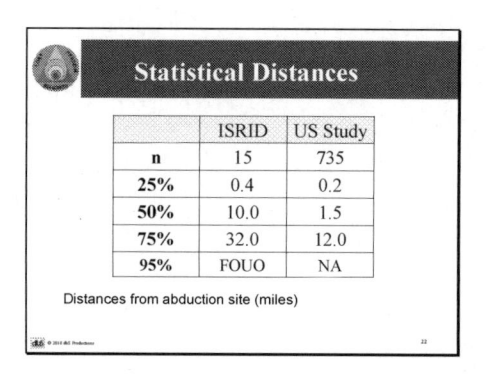

Statistical Distances

	ISRID	US Study
n	15	735
25%	0.4	0.2
50%	10.0	1.5
75%	32.0	12.0
95%	FOUO	NA

Distances from abduction site (miles)

Major difference between the ISRID data and US Study is that all ISRID data involved searches and abductions. Some subjects were found alive. The US Study with a much better statistical base only looked at stranger abductions that resulted in a homicide to a child. The 95% distance was removed from the book at the request of the NCMEC and is not publically available. In the Brown et al study (US Study) the data collection method precluded calculating the 95%. Additional statistics are available from the UK. Some of which are printed in Skip Stoffel's *Managing Land Search Operations* text book. Other critical statistical information can be found in the book *Geographic Profiling* by D. Kim Rossmo.

Unit 7 – Subject Categories

Slide 23

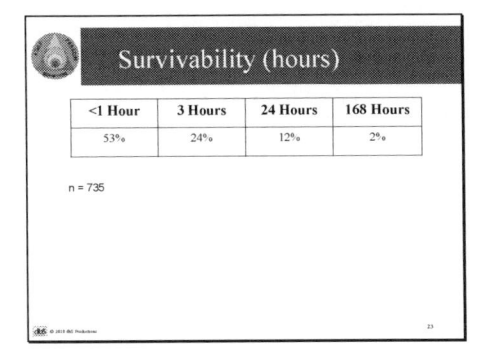

Survivability (hours)

<1 Hour	3 Hours	24 Hours	168 Hours
53%	24%	12%	2%

n = 735

Don't give up hope – there is only a 2% chance that a missing child was the victim of a stranger abduction to begin with – and even if they are, there is still a 26% chance they are still alive after you've been called!

Slide 24

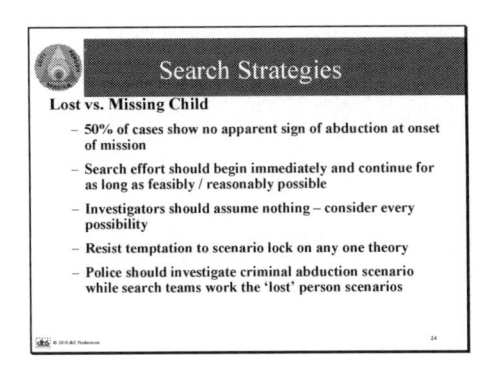

Search Strategies

Lost vs. Missing Child
- 50% of cases show no apparent sign of abduction at onset of mission
- Search effort should begin immediately and continue for as long as feasibly / reasonably possible
- Investigators should assume nothing – consider every possibility
- Resist temptation to scenario lock on any one theory
- Police should investigate criminal abduction scenario while search teams work the 'lost' person scenarios

The line between lost and missing is a very fat and fuzzy line. If the police suspect foul play, they are typically very reluctant to bring in outside resources to aid in the search since it is then considered a potential criminal investigation. Unfortunately, this cultural mentality is actually detrimental to the location and recovery of the victim. Specifically, the size and scope of the search area will undoubtedly exceed the capability of the local law enforcement agency. Culturally, the police will want to maintain control and may be concerned that the crime scene might be compromised if they do not find it first. However, given that most homicide victims are located by a casual observer anyway – there is absolutely no evidence that a 'non-police find' has ever prevented the police in bringing the accused to court and winning a conviction.

Unit 7 – Subject Categories

Slide 25

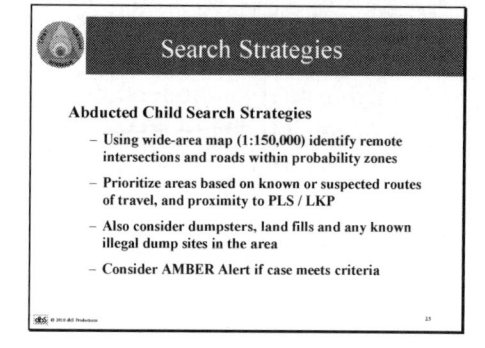

Search Strategies

Abducted Child Search Strategies
- Using wide-area map (1:150,000) identify remote intersections and roads within probability zones
- Prioritize areas based on known or suspected routes of travel, and proximity to PLS / LKP
- Also consider dumpsters, land fills and any known illegal dump sites in the area
- Consider AMBER Alert if case meets criteria

It is paramount the search manager consider the maximum potential search area – even if this means the search will include surrounding jurisdictions. Broad cooperation is a key element to success. A wide-area map must be utilized to identify potential body disposal sites. Initial efforts should have a strong focus on the 50% probability zone, as that will also include most 'lost' scenario strategies as well.

Slide 26

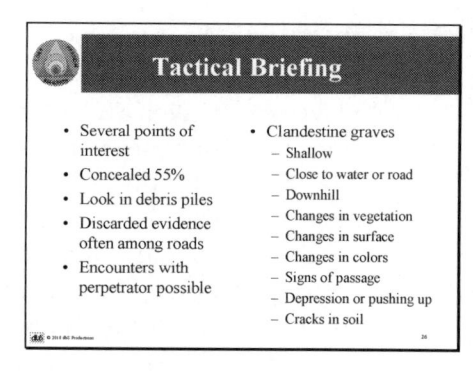

Tactical Briefing

- Several points of interest
- Concealed 55%
- Look in debris piles
- Discarded evidence often among roads
- Encounters with perpetrator possible
- Clandestine graves
  - Shallow
  - Close to water or road
  - Downhill
  - Changes in vegetation
  - Changes in surface
  - Changes in colors
  - Signs of passage
  - Depression or pushing up
  - Cracks in soil

Slide 27

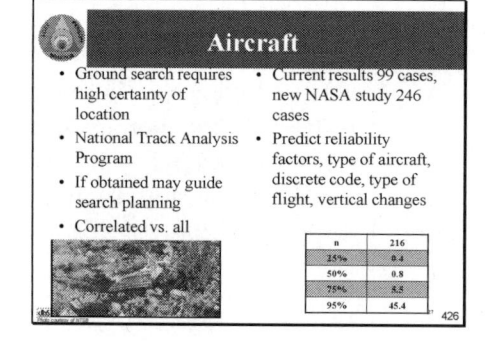

Aircraft

- Ground search requires high certainty of location
- National Track Analysis Program
- If obtained may guide search planning
- Correlated vs. all
- Current results 99 cases, new NASA study 246 cases
- Predict reliability factors, type of aircraft, discrete code, type of flight, vertical changes

n	216
25%	0.4
50%	0.8
75%	8.5
95%	45.4

Information presented in this PowerPoint presentation is more up-to-date than the information in the text book. At the time the textbook was printed only 99 cases had been examined. A second and third NASA study (also by Koester) later looked at 216 missing aircraft incidents that had radar data. All distances given in this PowerPoint are in nautical miles. The distances given in the text book are either statue miles or nautical miles. A copy of the NASA study is provided in the Instructor Resource DVD.

Unit 7 – Subject Categories

Slide 28

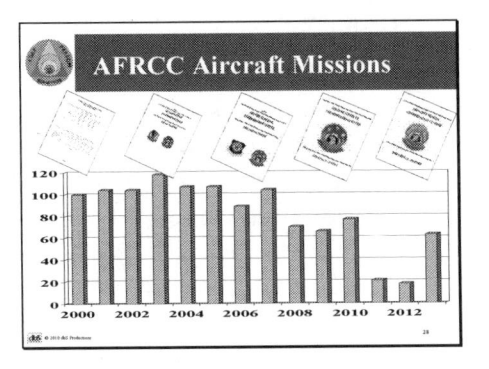

The dip in incidents in 2008 is most likely due to the poor economy and high price of AV Gas, which significantly reduced the amount of general aviation.

Slide 29

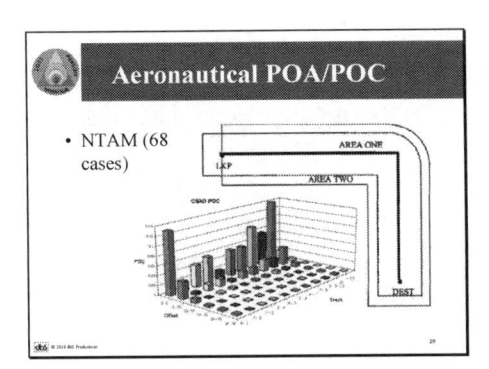

Previous statistical information on where to look or to base the POC/POA as taught by the USAF Inland SAR School was based upon the New Two Area Method (NTAM) which looked at 68 cases from Canada.

Slide 30

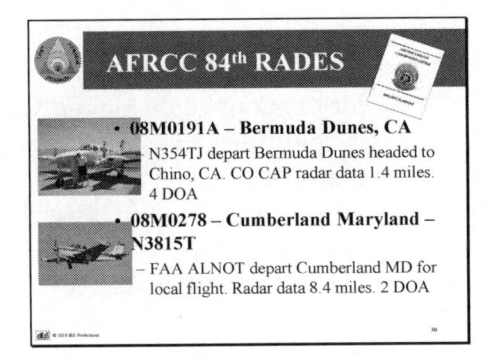

The most powerful tool used by the AFRCC for missing aircraft searches is radar data. The 2008 Annual Report (included in Instructor Resource DVD), describes several search incidents where radar data played a role. However, it is just a tool and not an absolute guide to where the aircraft is located.

Lost Person Behavior — September 2014

Unit 7 – Subject Categories

Slide 31

Animated Slide. Shows first percentage of incidents that typically have radar coverage. Second the question was asked to CAP pilots what is the relationship between the last radar plot and where the aircraft is found, finally the answer provided by the AFRCC to the same question. All answers were provided prior to the NASA study.

Slide 32

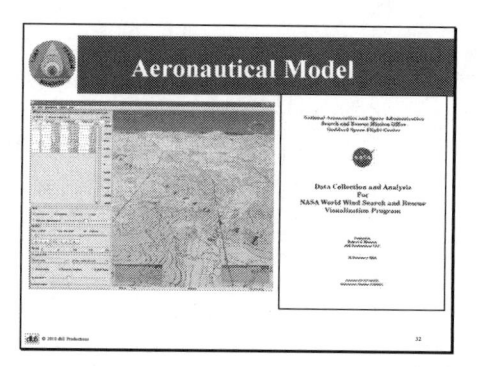

NASA is currently developing a software program called NASA World wind search and rescue prototype which has the goal of helping to better visualize where to look. As part of this project a study was conducted to collect basic statistics regarding missing aircraft.

Slide 33

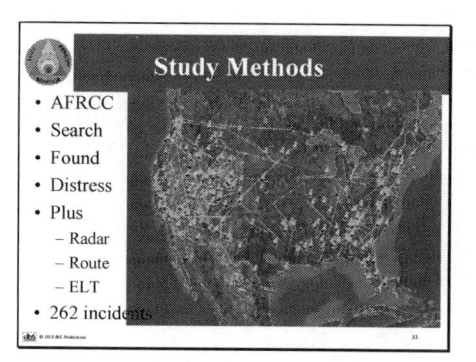

All incidents were taken from AFRCC records (which has an area of responsibility for the Contiguous United States). All incidents began as missing aircraft searches, the aircraft was eventually located, the cause of the incident was actual distress. In addition, information needed to be provided that included radar, route, or ELT data. The map shows the location of the missing aircraft (airplane icon), the last radar plot (yellow pushpin), and the route of the aircraft (white line). It can be noted the main clusters all occur in more mountainous areas except for Florida.

Lost Person Behavior

Unit 7 – Subject Categories

Slide 34

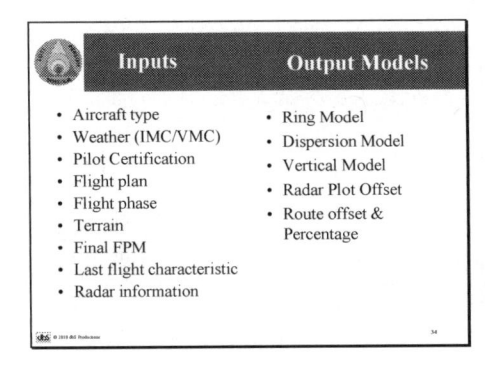

Inputs
- Aircraft type
- Weather (IMC/VMC)
- Pilot Certification
- Flight plan
- Flight phase
- Terrain
- Final FPM
- Last flight characteristic
- Radar information

Output Models
- Ring Model
- Dispersion Model
- Vertical Model
- Radar Plot Offset
- Route offset & Percentage

An input is information that may be known to a search planner prior to an aircraft being located. Some investigation may be required. An output is a model based upon case studies that gives probabilities of where the aircraft may be located.

Slide 35

Ring Model (Distance from Last Radar Plot in Nautical Miles)

Stat	All	In radar	Jet	VFR	Hook
Count	216	211	12	42	34
25%	0.4	0.4	0.6	0.5	0.3
50%	0.8	0.8	1.5	3.0	0.4
75%	5.5	4.5	2.4	13.1	1.5
95%	45.4	37.7	4.2	55.8	6.3

The actual study has a much more extensive collection of tables. This table illustrates some of the important differences between types of Aircraft, flight plans, and final flight characteristics.

Unit 7 – Subject Categories

Slide 36

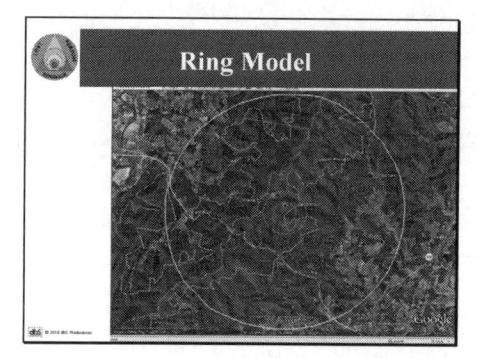

The ring model is the classic distance from the last radar plot, measured in crowflight distance. In this illustration of an actual incident the red ring is the 25% distance, the orange ring the 50% distance, and the yellow ring the 75%.

Slide 37

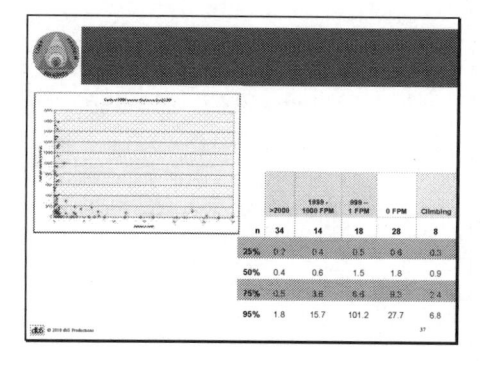

Slide 38

Snake River, ID – Shelton, WA
N430A
Cessna 208B – Grand Caravan

Lost Person Behavior — 130 — September 2014

Unit 7 – Subject Categories

Slide 39

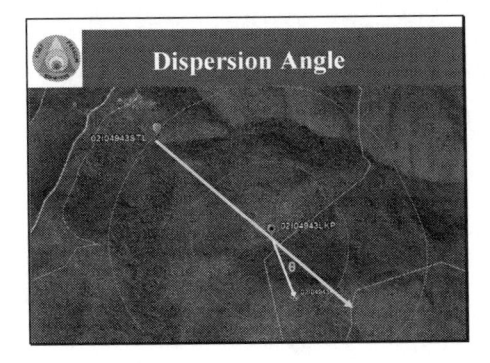

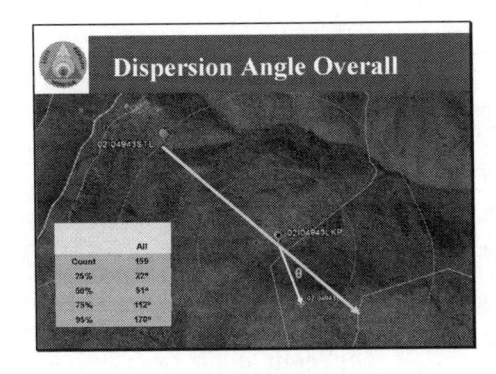

Slide 40

Slide 41

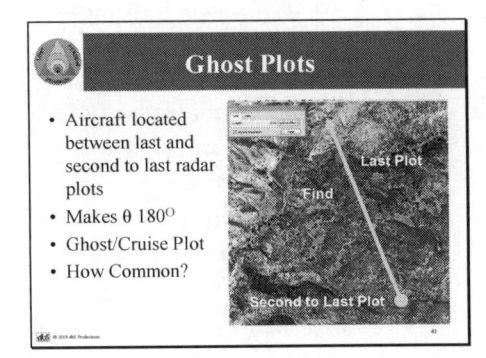

Ghost plots are more likely when data from FAA radars. It is artifact of how the software handles the information. Right now, it is unknown how common they are. On more recent data supplied by the AFRCC in some cases the last plot was labeled as a cruise plot.

Lost Person Behavior

Unit 7 – Subject Categories

Slide 42

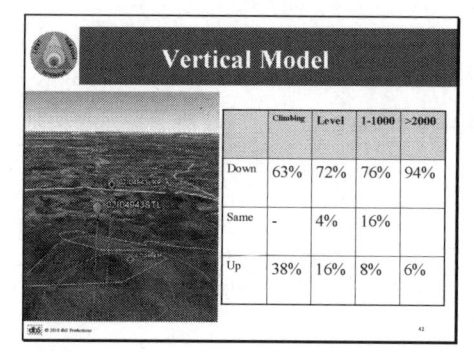

Vertical Model

	Climbing	Level	1-1000	>2000
Down	63%	72%	76%	94%
Same	-	4%	16%	
Up	38%	16%	8%	6%

Slide 43

Page involves several clicks. The first click shows the flight path of a Cessna 172 that departed from California and was traveling east across the desert. The second click zooms in somewhat to show the last 9 radar hits or last minute and half of flight. The next frame show the last three radar hits and adds a blue color showing the altitude to clearly show elevation above the flight level. Aircraft continued flying straight and level. Students should be asked what they project might be the location of the aircraft. The final frame show what the lighting was at the actual time of the last three radar hits. The next slide will reveal the find location.

This slide can also be used to discuss where the altitude information comes from. The transponder when it return radar data sends the altitude from the aircrafts altimeter. If the altimeter is incorrect then the altitude will be incorrect. This can be caused by malfunctions or by changes in barometric pressure. In addition, the altitude is only reported in increments of 100 feet.

Lost Person Behavior — September 2014

Unit 7 – Subject Categories

Slide 44

Find location shows the pilot did pull up some at the last moment.

Slide 45

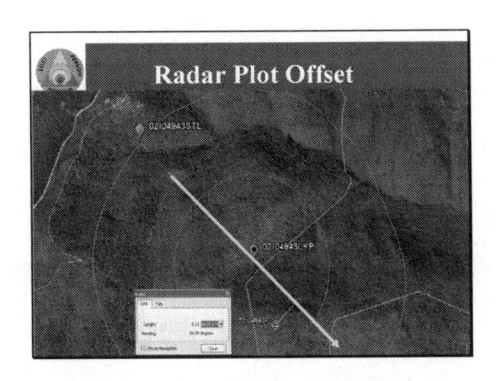

Slide 46

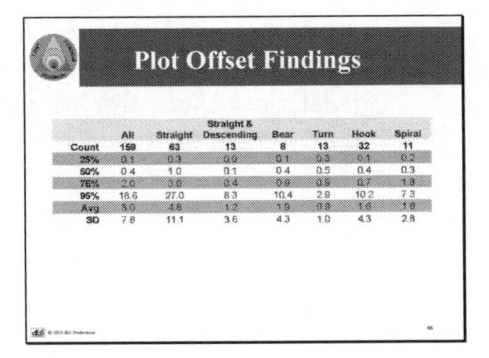

Lost Person Behavior — 133 — September 2014

Unit 7 – Subject Categories

Slide 47

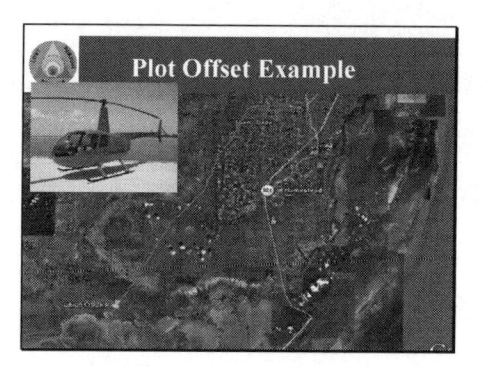

Plot Offset Example

Helicopter takes off from Miami on its way to the Keys. Last radar plot it over the everglades. Should ground teams be deployed?

Slide 48

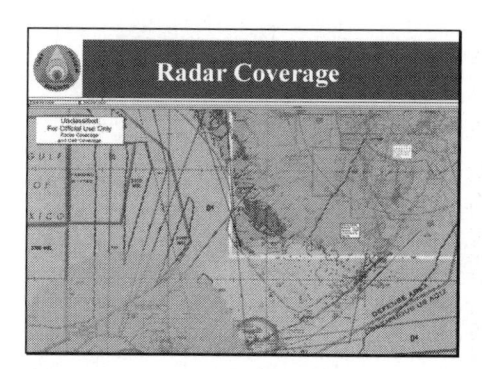

Radar Coverage

It is important to get an idea about the end of radar coverage. When the last plot is at the end of coverage it is more suspect. Must ask the AFRCC to provide radar coverage maps. They are not classified information but they are FOUO. They can be shared with anyone who has the need to know.

Slide 49

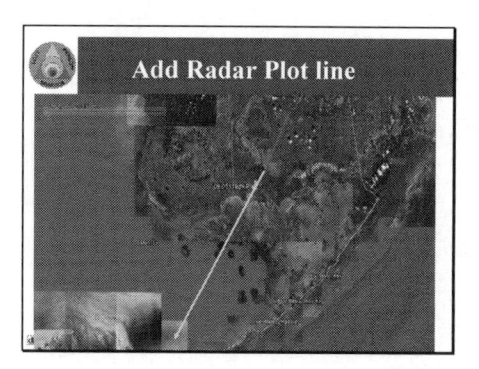

Add Radar Plot line

Based upon cell phone forensics the flight appeared to have continued after the last radar plot. Therefore, flying the route first is the best initial action.

Lost Person Behavior — September 2014

Unit 7 – Subject Categories

Slide 50

Find Location

Find was made by the CAP. Rescue was done by helicopter

Slide 51

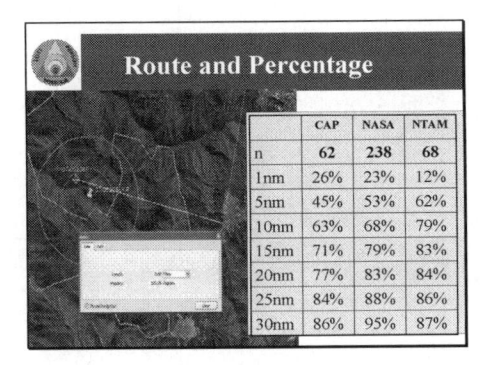

Route and Percentage

	CAP	NASA	NTAM
n	62	238	68
1nm	26%	23%	12%
5nm	45%	53%	62%
10nm	63%	68%	79%
15nm	71%	79%	83%
20nm	77%	83%	84%
25nm	84%	88%	86%
30nm	86%	95%	87%

Slide 52

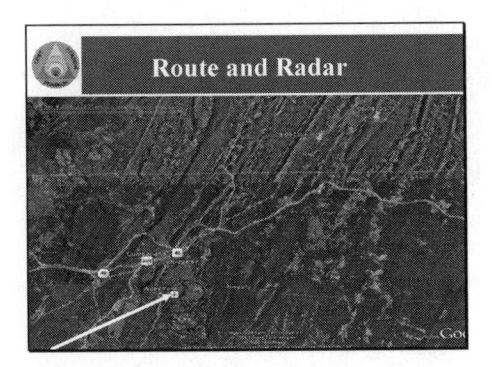

Route and Radar

Lost Person Behavior   September 2014

Unit 7 – Subject Categories

Slide 53

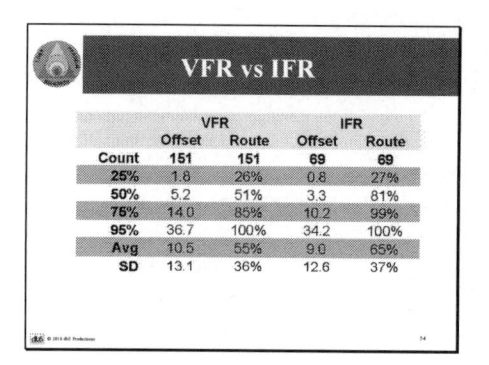

Find Location

Slide 54

VFR vs IFR

	VFR		IFR	
	Offset	Route	Offset	Route
Count	151	151	69	69
25%	1.8	26%	0.8	27%
50%	5.2	51%	3.3	81%
75%	14.0	85%	10.2	99%
95%	36.7	100%	34.2	100%
Avg	10.5	55%	9.0	65%
SD	13.1	36%	12.6	37%

Visual Flight Rules (VFR) versus IFR (Instrument Flight Rules) give very different results. VFR crashes seem to be distributed evenly along the flight route. IFR crashes appear more associated with the landing, with 25% of the flights crashing in the last 1% of the route.

Slide 55

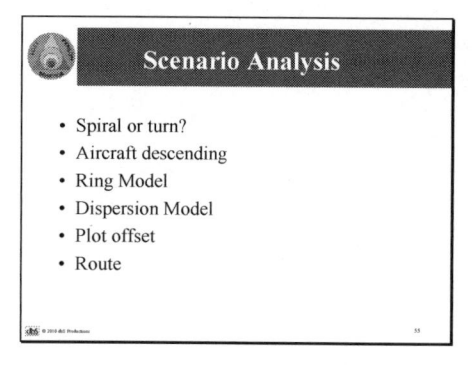

Scenario Analysis

- Spiral or turn?
- Aircraft descending
- Ring Model
- Dispersion Model
- Plot offset
- Route

An aircraft incident results in many different models to consider.

Lost Person Behavior   September 2014

Unit 7 – Subject Categories

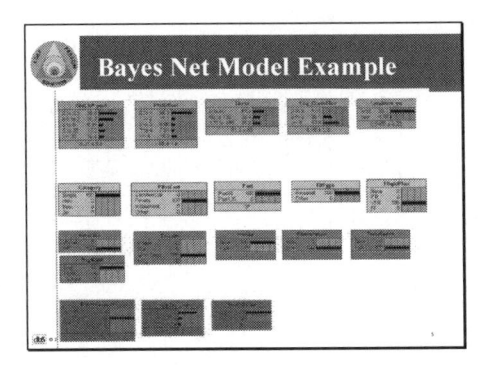

Slide 56

Example of a Bayes net that automatically updates probability has more factors are known.

Slide 57

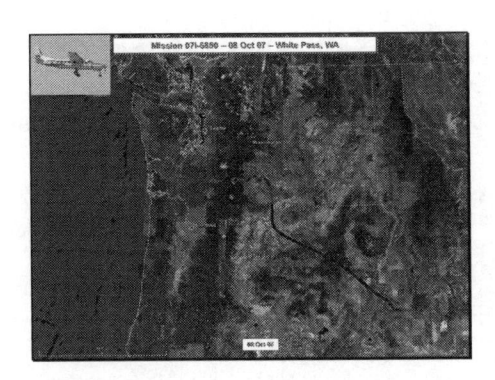

Snake River, ID – Shelton, WA N430A
Cessna 208B – Grand Caravan

Slide 58

Snake River, ID – Shelton, WA
N430A
Cessna 208B – Grand Caravan

Lost Person Behavior — September 2014

Unit 7 – Subject Categories

Slide 59

Snake River, ID – Shelton, WA
N430A
Cessna 208B – Grand Caravan

Slide 60

Snake River, ID – Shelton, WA
N430A
Cessna 208B – Grand Caravan

Slide 61

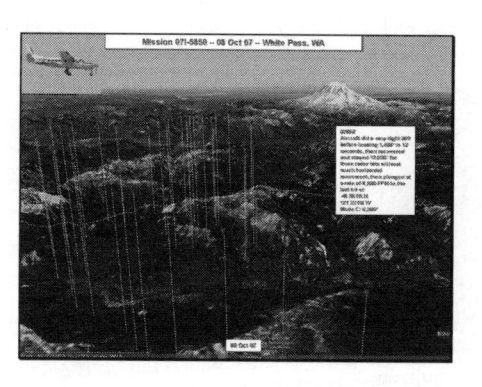

Snake River, ID – Shelton, WA
N430A
Cessna 208B – Grand Caravan

Lost Person Behavior   September 2014

Unit 7 – Subject Categories

Slide 62

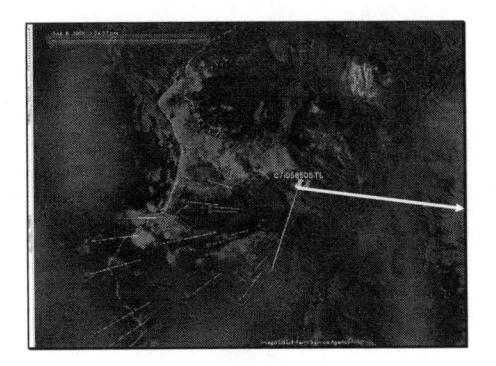

Looking at last plots from top view, white arrow is drawn from secont to last (STL) to last radar plot.

Slide 63

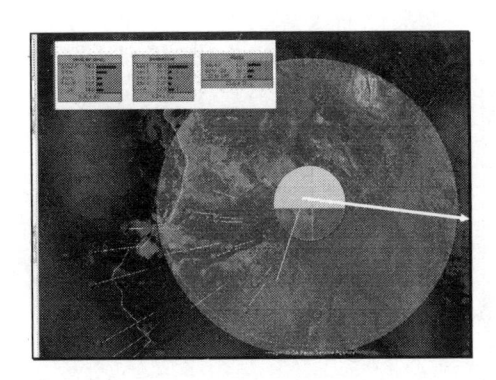

Using the Bayesian Model for the specifics of this incident the distance from the LKP within a 0.5 nm ring is 39.4 percent. From 0.5 to 2. nm is 22% for a total of 61.4% within 2.0 nm of the last radar plot.

Slide 64

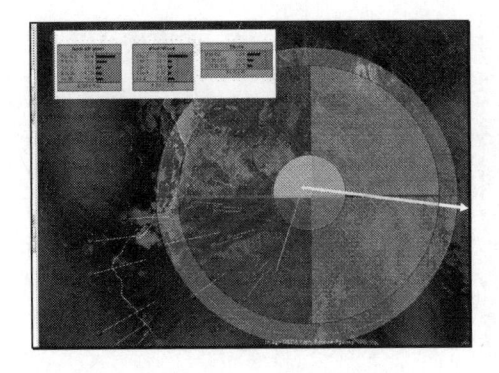

However, the original model did not look at the effect of vertical Feet per minute. Since the aircraft was descending more than 2000 feet per minute we can add that 95% ring at 1.8 nm.

Lost Person Behavior   September 2014

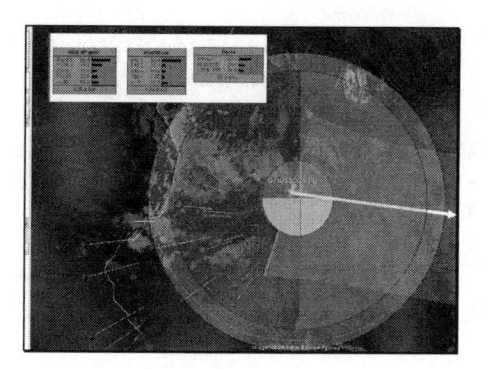

Slide 65 — The plot offset of 1 nm is shown by the white box. There is a 53.4% chance of the aircraft being within this box.

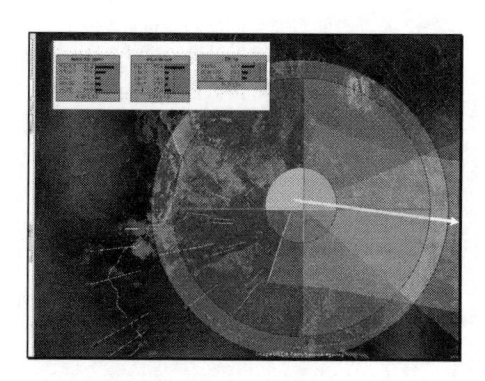

Slide 66 — The Theta angle of 60 degrees is shown. The aircraft has a 52% chance of being within this area. After all three models are displayed, Instructor should ask class where the likely areas to look are from a ground perspective. It would include everything within the 25% ring, and the cone where the three models overlap.

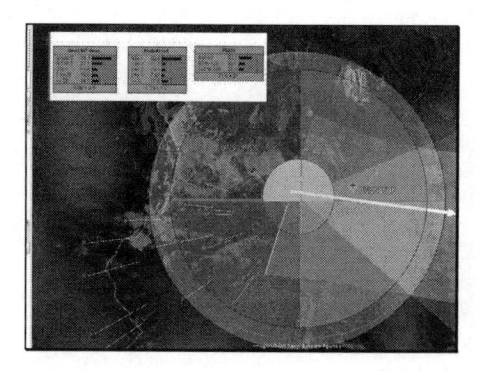

Slide 67 — Slide shows the actual location of the aircraft.

Unit 7 – Subject Categories

Slide 68

Slide 69

Initial Assignment

- Cherokee Six single engine (285 HP Lancer) w/ five souls on board departed Manteo, NC enroute to Ohio on 16 Aug.
- Reported missing 18 Aug.
- Sufficient information to launch a ground search?
- What actions can be taken?

Additional details can be found in the Instructor's Guide. Activity 7-2

Slide 70

- NTAP obtained 23 Aug after investigation.
- Pilot flying VFR, no flight plan, ceiling 2000'
- Determine high probability area, determine ground search area

Lost Person Behavior

Unit 7 – Subject Categories

Slide 71

In most cases, it can be pointed out to the students that the last plot was a ghost plot. The factors that help determine the potential of a ghost plot included, straight line projection, same altitude, and most importantly it would have required a dramatic increase in airspeed along with gaining sufficient altitude to cross over the mountain ridge. In a class of experienced pilots it might be best to allow them to make that determination. It will typically add 10-15 minutes to the time if they are not told about the ghost plot.

Slide 72

Slide 73

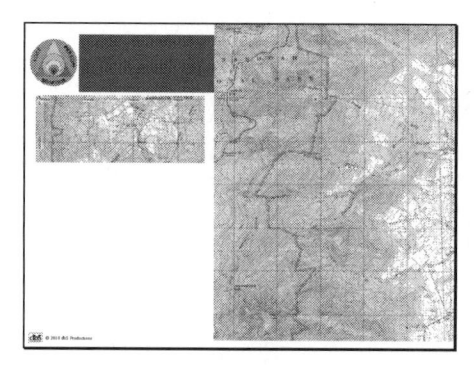

Lost Person Behavior   September 2014

Unit 7 – Subject Categories

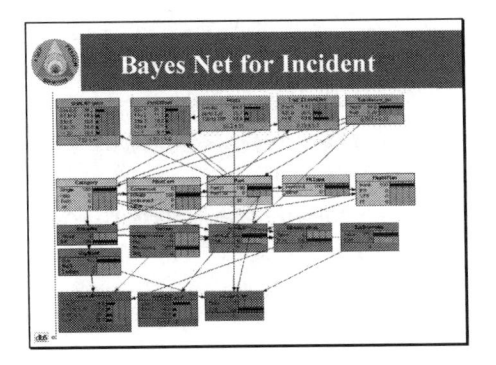

Slide 74

Bayes net example specific to this case. In most cases this can be skipped.

Slide 75

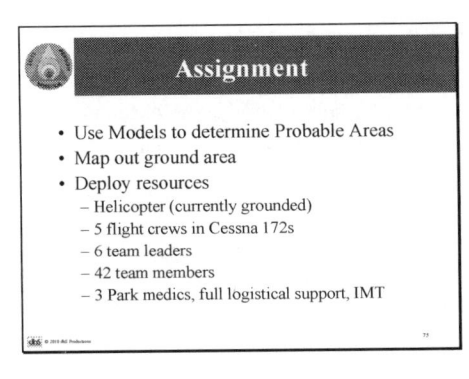

Slide slightly updated to better reflect actual search resources.

Slide 76

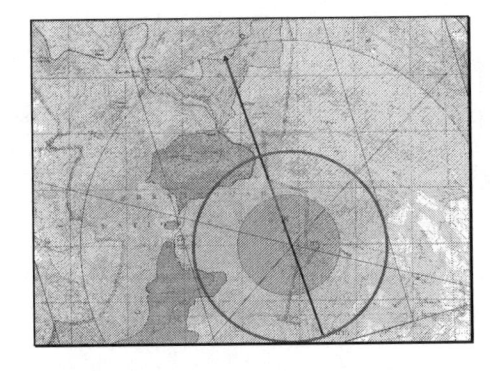

The time stamp of 17:45:01 represents the last radar plot which also reported an altitude of 2400 feet. The black arrow is drawn from the second to last radar plot through the last one and continues. Using the Bayesian model (shown on previous PowerPoint slide) the orange area around the last plot has a 38.2 percent chance of containing the aircraft, the yellow circle goes out 2 nm and has a 57% chance of containing the aircraft. The red circle represents where the plan would be expected to be in twelve seconds which represents the interval of radar reported positions. The plot offset shown by the green box of 1 nm has a 51% chance of containing the aircraft. The angle of

Unit 7 – Subject Categories

dispersion of 60 degrees has a 64% chance of containing the aircraft. The orange areas represent the 2400 contour interval. This terrain would be higher than the elevation the aircraft was flying at. This slide may be shown to the class if the participants

Slide 77

Once the ground fog lifted and it was possible to launch the helicopter the find came rather quickly. The area of dead vegetation was easy to spot from the air.

Slide 78

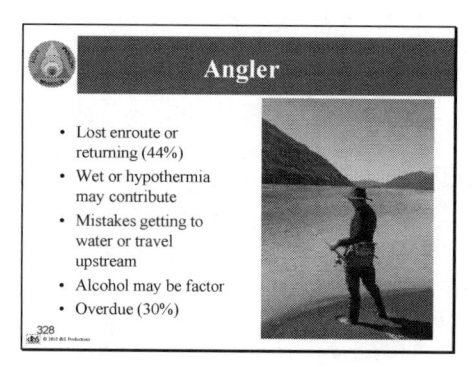

Angler
- Lost enroute or returning (44%)
- Wet or hypothermia may contribute
- Mistakes getting to water or travel upstream
- Alcohol may be factor
- Overdue (30%)

Text Reference: Page 106-110

Unit 7 – Subject Categories

Slide 79

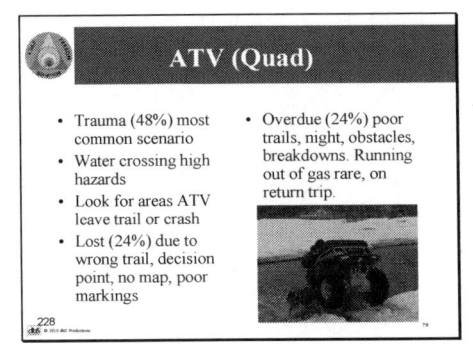

ATV (Quad)
- Trauma (48%) most common scenario
- Water crossing high hazards
- Look for areas ATV leave trail or crash
- Lost (24%) due to wrong trail, decision point, no map, poor markings
- Overdue (24%) poor trails, night, obstacles, breakdowns. Running out of gas rare, on return trip.

Text Reference: Page 111-114

Slide 80

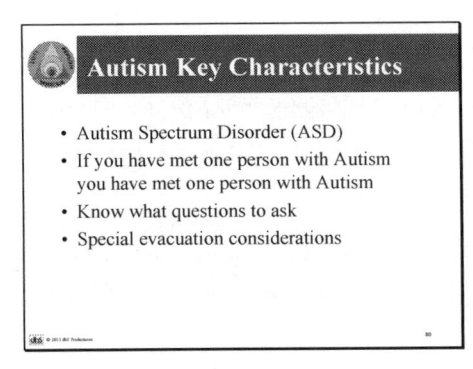

Autism Key Characteristics
- Autism Spectrum Disorder (ASD)
- If you have met one person with Autism you have met one person with Autism
- Know what questions to ask
- Special evacuation considerations

Slide 81

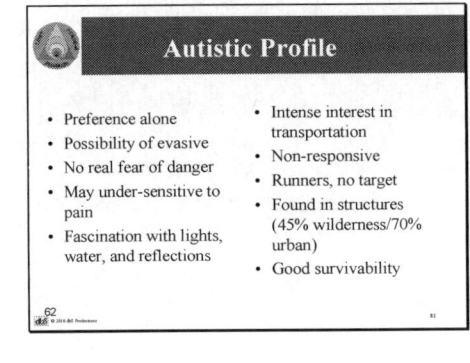

Autistic Profile
- Preference alone
- Possibility of evasive
- No real fear of danger
- May under-sensitive to pain
- Fascination with lights, water, and reflections
- Intense interest in transportation
- Non-responsive
- Runners, no target
- Found in structures (45% wilderness/70% urban)
- Good survivability

Text Reference: 115-119 Instructors DVD contains seven additional references for background material on Autism and spectrum disorders.

Lost Person Behavior — September 2014

Slide 82

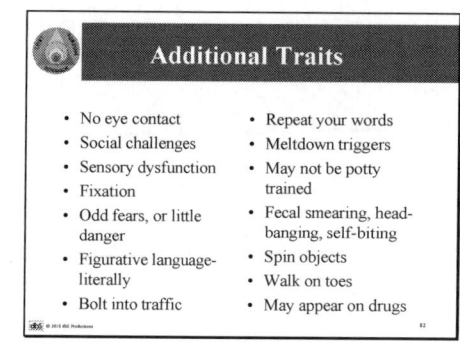

Additional Traits

- No eye contact
- Social challenges
- Sensory dysfunction
- Fixation
- Odd fears, or little danger
- Figurative language- literally
- Bolt into traffic
- Repeat your words
- Meltdown triggers
- May not be potty trained
- Fecal smearing, head-banging, self-biting
- Spin objects
- Walk on toes
- May appear on drugs

Slide 83

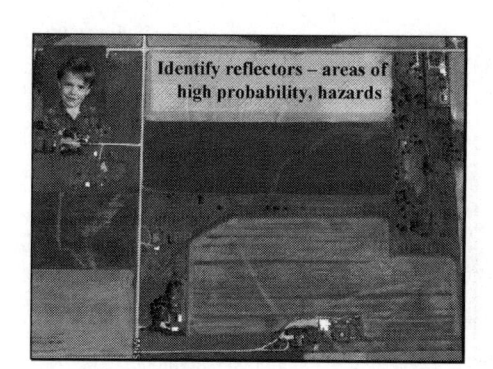

Identify reflectors – areas of high probability, hazards

Mason, a five-year with Autism escaped out of his sister's window. The older sister called her mother who immediately called 911. Police arrived at the house within minutes. What are the reflectors or points of high probability or hazard? Have the students point them out. He was found within 25 minutes in the pond just under his picture by his mother. She had told police to check the pond, but the pond was not visible from the road. It is believed he may have been attracted to the pond by a nearby windmill (can be seen if you use Google Earth ground level view). CPR revived him, but he died two days later in the hospital. Drowning is the number one cause of death among those with Autism who go missing.

http://www.masonallenmedlamfoundation.webs.com/ for more information on the incident.

Slide 84

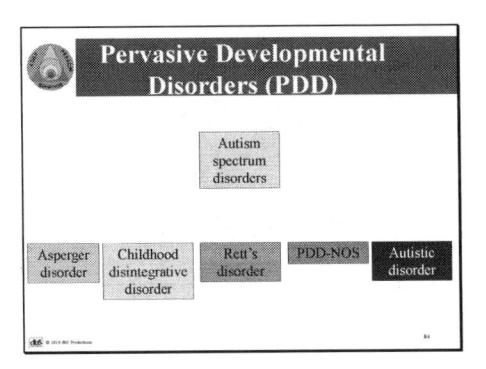

Pervasive Developmental Disorders (PDD)
- Autism spectrum disorders
- Asperger disorder
- Childhood disintegrative disorder
- Rett's disorder
- PDD-NOS
- Autistic disorder

With the revised version of DSM V, Asperger's disorder is no longer a term used. However, many clinical professionals disagree with this, so expect to hear the term in the future. Instead all of these terms will simply be termed Autism Spectrum disorders.

Slide 85

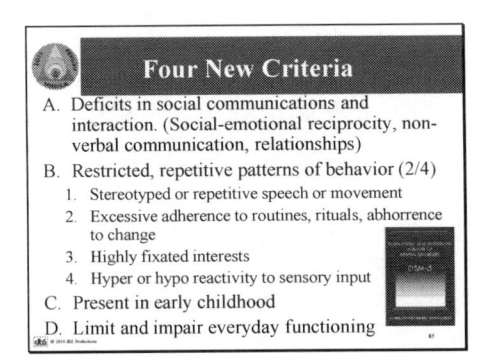

Four New Criteria
A. Deficits in social communications and interaction. (Social-emotional reciprocity, non-verbal communication, relationships)
B. Restricted, repetitive patterns of behavior (2/4)
 1. Stereotyped or repetitive speech or movement
 2. Excessive adherence to routines, rituals, abhorrence to change
 3. Highly fixated interests
 4. Hyper or hypo reactivity to sensory input
C. Present in early childhood
D. Limit and impair everyday functioning

The revised DSM V criteria for Autism Spectrum Disorders has four requirements. The full criteria can be seen online at http://www.dsm5.org/ProposedRevisions/Pages/proposedrevision.aspx?rid=94 The changes became formal on May 2013. A single term will apply, simply Autism Spectrum Disorder or ASD.

Slide 86

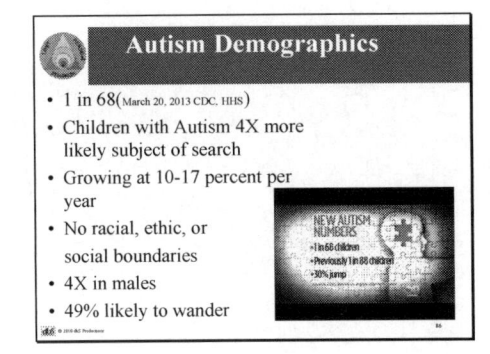

Autism Demographics
- 1 in 68 (March 20, 2013 CDC, HHS)
- Children with Autism 4X more likely subject of search
- Growing at 10-17 percent per year
- No racial, ethic, or social boundaries
- 4X in males
- 49% likely to wander

The additional Blue curve is because the formal definition of what is Autism changed in 2000. This caused a slight increase in the number of cases. According to a MMWR from the Center for Disease Control (CDC) for every 110 8-year olds (peak age for reporting) one would be expected to have an Autism Spectrum Disorder (ASD). This works out to 0.9 percent. The previous prevalence was 1 in 178 then down to 1 in 150. The full PDF is included in the instructor's DVD. The ISRID database contains 1669 normal children. When this number is multiply by 0.9 percent the result is 15. Since the ISRID database contains 62 cases of searches for Autism then it can be concluded that

Unit 7 – Subject Categories

something about Autism itself greatly increases the chances of becoming the subject of a reported search. A one minute sound clip from the CDC is included that discusses the rate and Autism. The sound clip needs to be in the same directory as the PPT presentation. The file name is mmwr1_010710.mp3

Slide 87

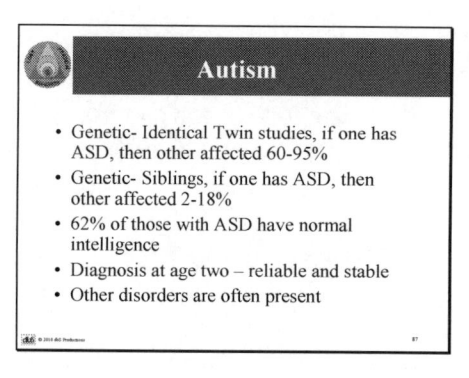

Autism
- Genetic- Identical Twin studies, if one has ASD, then other affected 60-95%
- Genetic- Siblings, if one has ASD, then other affected 2-18%
- 62% of those with ASD have normal intelligence
- Diagnosis at age two – reliable and stable
- Other disorders are often present

Slide 88

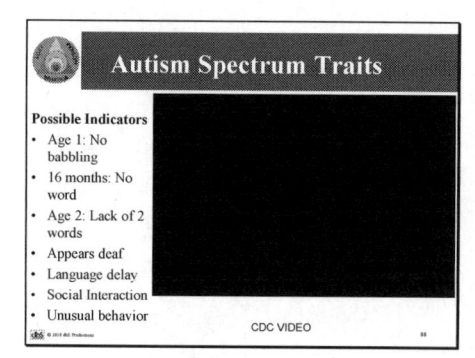

Autism Spectrum Traits

Possible Indicators
- Age 1: No babbling
- 16 months: No word
- Age 2: Lack of 2 words
- Appears deaf
- Language delay
- Social Interaction
- Unusual behavior

CDC VIDEO

Lost Person Behavior

Unit 7 – Subject Categories

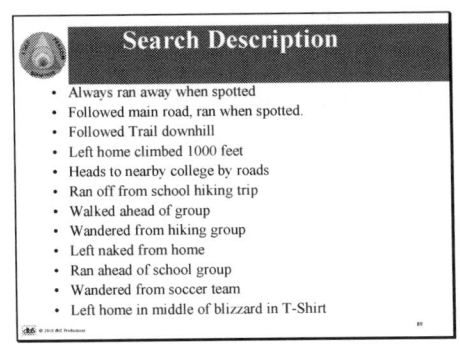

Slide 89

Search Description
- Always ran away when spotted
- Followed main road, ran when spotted.
- Followed Trail downhill
- Left home climbed 1000 feet
- Heads to nearby college by roads
- Ran off from school hiking trip
- Walked ahead of group
- Wandered from hiking group
- Left naked from home
- Ran ahead of school group
- Wandered from soccer team
- Left home in middle of blizzard in T-Shirt

Slide 90

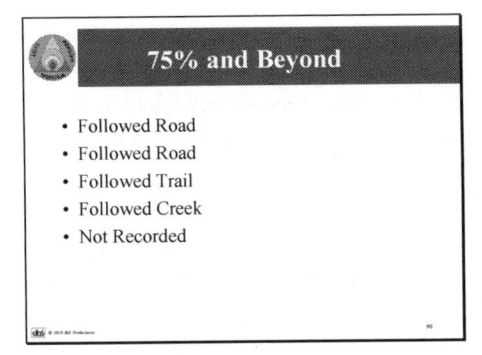

75% and Beyond
- Followed Road
- Followed Road
- Followed Trail
- Followed Creek
- Not Recorded

Featured followed by those who traveled beyond the statistical 75% ring.

Slide 91

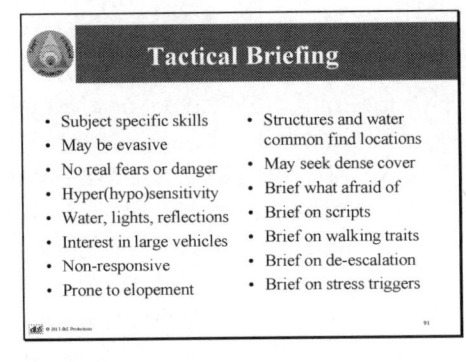

Tactical Briefing
- Subject specific skills
- May be evasive
- No real fears or danger
- Hyper(hypo)sensitivity
- Water, lights, reflections
- Interest in large vehicles
- Non-responsive
- Prone to elopement
- Structures and water common find locations
- May seek dense cover
- Brief what afraid of
- Brief on scripts
- Brief on walking traits
- Brief on de-escalation
- Brief on stress triggers

Unit 7 – Subject Categories

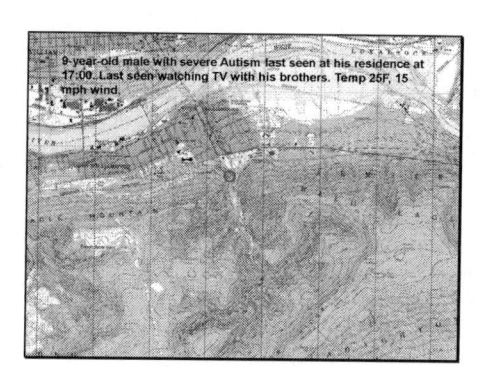

Slide 92

Scenario: A 9-year-old male with severe Autism is last seen in his residence at 5 P.M. on December 18. He was watching television with his siblings. He was last seen wearing a long-sleeved shirt, blue jeans, and slippers. He was not wearing a coat or hat. The temperature is currently 25 F (-4 C). Conditions are clear with a 15 mph (24 kmph) wind. More details provided in Instructor's Activity Guide Activity 7-3

Slide 93

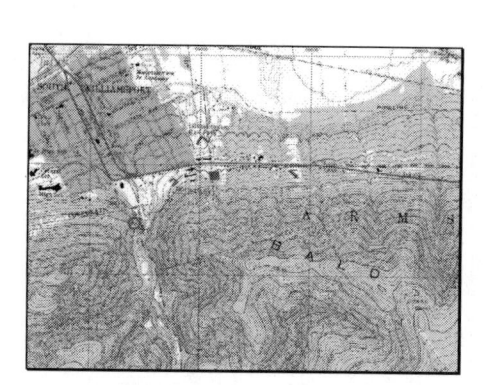

Instructor's Notes. On December 18 a 9-year-old male with Autism was last seen at his residence at 5 p.m. He was watching television with his siblings. He was wearing a long-sleeved shirt, blue jeans, and slippers. He was not wearing a coat or hat. The following night temperatures dropped 15 degrees F, the following day temperatures remained below 20 degrees F. The search involved over 350 personnel with a large percent of emergent volunteers and a substantial non-coordinated volunteers. Must of the initial efforts were directed to the town of South Williamsport itself where a sighting had placed him at the local McDonalds. Since the sighting was made by the son of the Chief of Police it was given more weight than might otherwise occur.

The subject had severe autism, was non-verbal, described by parents has hyper-active. He was attracted to lights. The subject was found deceased (the only fatality currently in the database – a second fatality from Canada will go into the database for 2009.)
The instructor should present the basic scenario and ask the participants what additional investigative information they would like and also what actions they would like to take. As with most map scenarios "What do you want to know,

Lost Person Behavior

Unit 7 – Subject Categories

where do you want to go." The key points are being attracted to lights and students doing a structure search. He was found after 58 hours of searching 12 meters away from the radio towers. He was found dead by a searcher doing independent searching (i.e. not part of the formal search effort).

Slide 94

Google Earth view, showing pushpins where the radio towers are located.

Slide 95

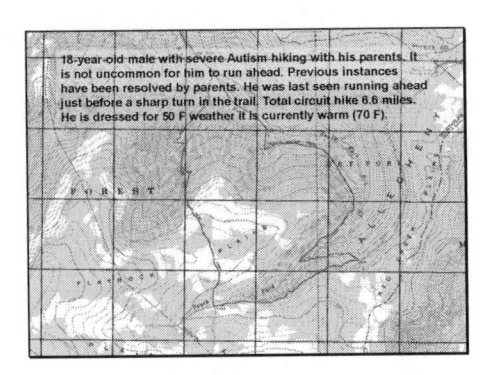

On Sunday October 14 an 18-year-old male with severe Autism went hiking with his parents in the Dolly Sods Wilderness Area in West Virginia. He is unable to communicate verbally and has a mental age of a 3-4 year old child. While he resides in a nearby city he frequently hikes with his family. It is not uncommon for him to run ahead and he has been missing for short periods before that the family was always able to resolve. The intended route was to start at the trail head on Forest Service Road 19 (not marked on the topographic maps but listed on Google Earth software). They took the Boar's Nest Trail (2.53 miles) South climbing to the Flatrock Plains. They then intended to take the South Fork Fire

Lost Person Behavior

Unit 7 – Subject Categories

Road (FR70) east (1.34 miles) to the South Prong Trail. They would then follow the South Prong Trail Northwest for 3.73 miles back to the trailhead. The total circuit hike would have been 6.6 miles. The weather is unseasonably warm (70 degree F highs). He does not carrying any supplies and is dressed for 50 degree weather.

Slide 96

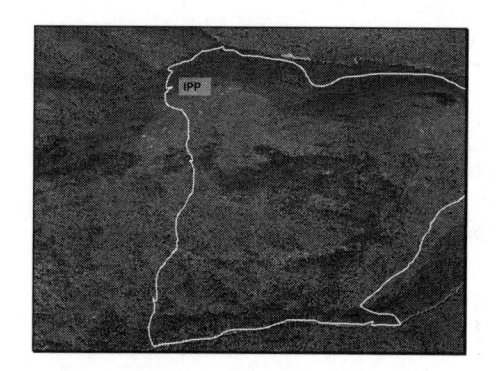

Google map view of circuit hike.

Slide 97

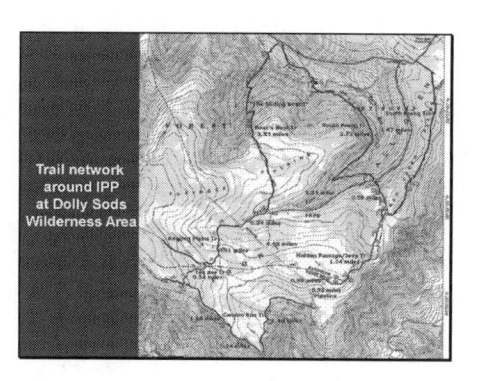

Trail network map.

Lost Person Behavior

Unit 7 – Subject Categories

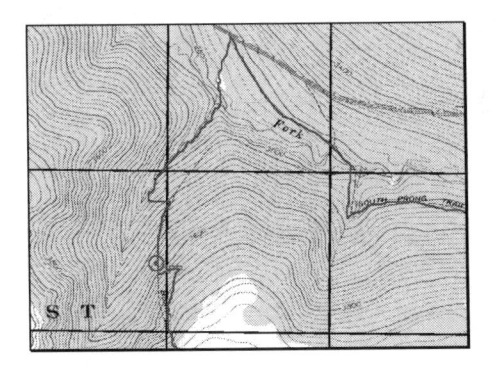

Slide 98

Close up view of IPP, notice the sharp switchback did not appear on the old USGS map, but was visible once the trail update was downloaded.

Slide 99

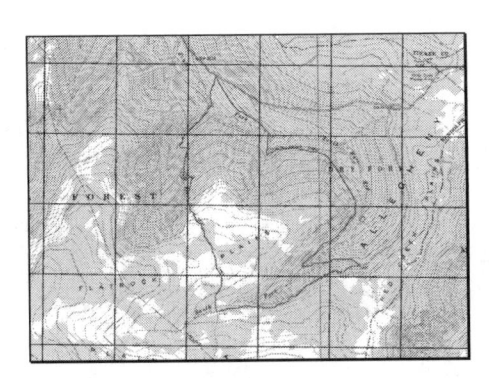

This slide is designed to be used for the actual deployment of tasks or after presentation from the class. The slide first shows a repeat of the general route, this is done to hide the find location and clue if the instructor elects to pass out slide handouts. The first click removes the cover (looks like nothing happens. The second click will reveal the location of the hat. The third click will reveal the find location. The subject was found alive after being missing for four days.

Slide 100

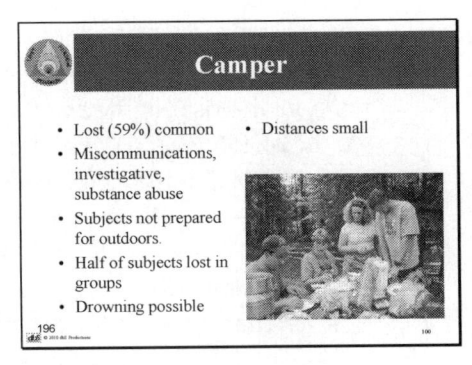

Text reference: 120-124

Lost Person Behavior — 153 — September 2014

Unit 7 – Subject Categories

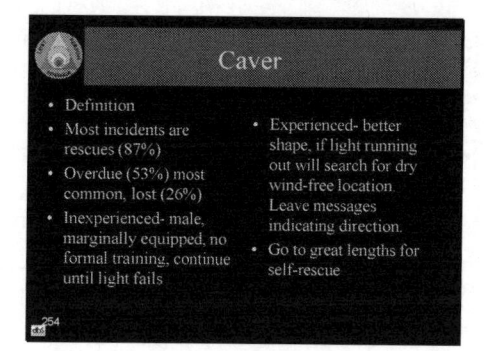

Slide 101

Caver
- Definition
- Most incidents are rescues (87%)
- Overdue (53%) most common, lost (26%)
- Inexperienced- male, marginally equipped, no formal training, continue until light fails
- Experienced- better shape, if light running out will search for dry wind-free location. Leave messages indicating direction.
- Go to great lengths for self-rescue

Text Reference: 123-128

Slide 102

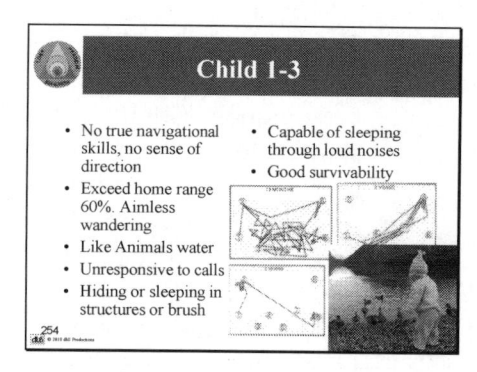

Child 1-3
- No true navigational skills, no sense of direction
- Exceed home range 60%. Aimless wandering
- Like Animals water
- Unresponsive to calls
- Hiding or sleeping in structures or brush
- Capable of sleeping through loud noises
- Good survivability

Text reference: 129-133 The three illustrations are from a study that showed the movement of different age children placed in a nursery school setting with eight different toys in the room. The time period shown is 7-minutes. Taken from the book *Your Two-Year-Old* by Louise Bates Ames and Frances Ilg.

Slide 103

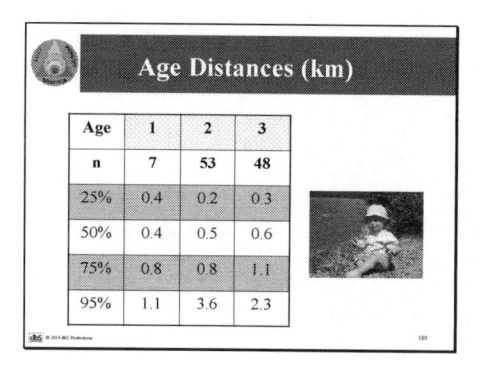

Age Distances (km)

Age	1	2	3
n	7	53	48
25%	0.4	0.2	0.3
50%	0.4	0.5	0.6
75%	0.8	0.8	1.1
95%	1.1	3.6	2.3

Distance from the IPP (only in Kilometers) broken down my specific age. The data for one-year olds is limited and the results are somewhat suspect. With more data I would imagine the 25% may decrease. Most one-year old searches are resolved by family, friends, and local law enforcement without resorting to a full search that qualifies for the database. This data is not available in the text book.

Lost Person Behavior — 154 — September 2014

Unit 7 – Subject Categories

Slide 104

Tactical Briefing
- No concept of lost – no navigational skills
- Aimless wandering
- Attracted to animals and water
- Brief on reaction to dogs, uniforms, lights, attraction to water, roads
- Difficult to detect; sleeping, small, enter into small spaces, penetrate brush

Slide 105

Child 4-6
- More mobile and capable than 1-3
- Understand lost
- Attempt to return
- Mind mixture of reason, fantasy, and faulty logic
- Can be taught to turn around for landmarks
- Focused on endpoints neglect route
- Accompany peers distance from home
- Drawn to animals or water
- May not understand return trip when playing
- Hide in structures

Text reference: 134-138

Slide 106

Child 7-9
- Developed directional skills
- Able to read maps
- Have mental map but may be wrong
- 400% over home range
- When lost use trail road strategy
- May stay put, but like view of area
- Intentional hiding to avoid punishment
- Same fears as adults but greater
- Girls more likely to be inside, boys outside

Text Reference: 139-143

Lost Person Behavior

Unit 7 – Subject Categories

Slide 107

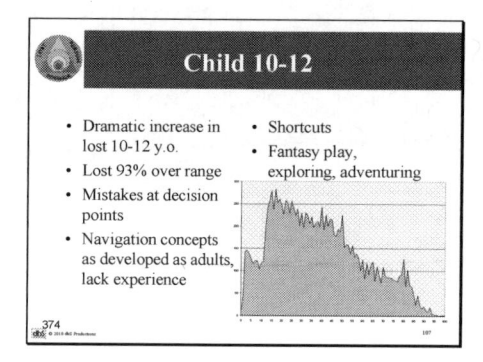

Child 10-12
- Dramatic increase in lost 10-12 y.o.
- Lost 93% over range
- Mistakes at decision points
- Navigation concepts as developed as adults, lack experience
- Shortcuts
- Fantasy play, exploring, adventuring

Text Reference: 144-149

Slide 108

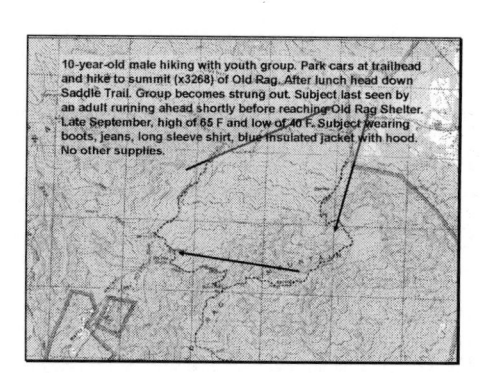

10-year-old male hiking with youth group. Park cars at trailhead and hike to summit (x3268) of Old Rag. After lunch head down Saddle Trail. Group becomes strung out. Subject last seen by an adult running ahead shortly before reaching Old Rag Shelter. Late September, high of 65 F and low of 40 F. Subject wearing boots, jeans, long sleeve shirt, blue insulated jacket with hood. No other supplies.

A 10 year old male was hiking with his youth group. They parked their cars at the Weakley Hollow Parking lot located at the boundary of Shenandoah National Park and indicated by the Gate. The group hiked the Ridge Trail to the Summit (x3268) of Old Rag Mountain where they stopped and had lunch. The group then headed down the Saddle Trail. The group then became strung out with several smaller groups of youths and adults. The ten year old was last seen by an adult running ahead shortly before reaching Old Rag Shelter. They were suppose to wait at the four-way junction (Saddle trail, Weakley Hollow Road, Berry Hollow Road, and Old Rag Road.)According to some other youths in the group the subject had been talking about how much fun it would be to go off trail. It is late September with day time's highs of 65 F and nighttime lows of 40 F. The subject is wearing hiking boots, jeans, a long sleeve shirt, and a blue insulated jacket with a hood. He is not carrying any supplies.

Unit 7 – Subject Categories

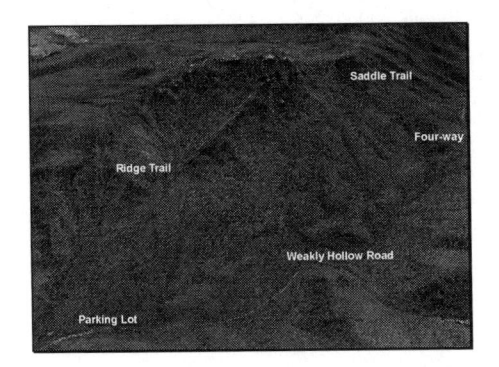

Slide 109　　　　　　　　　　　　　　　　　　　　　Google Earth image of circuit hiking looking to the South.

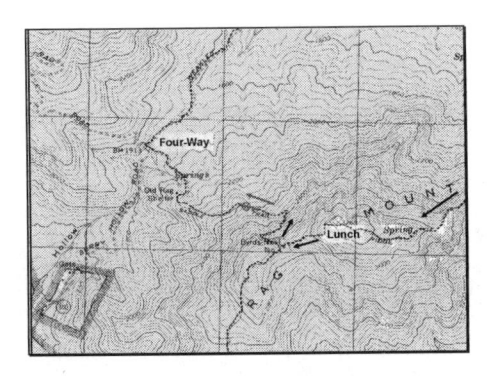

Slide 110　　　　　　　　　　　　　　　　　　　　　Detailed view of IPP

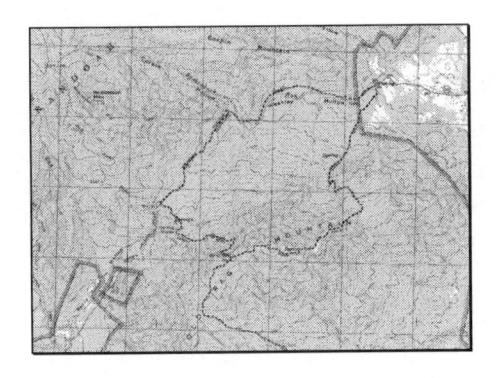

Slide 111　　　　　　　　　　　　　　　　　　　　　Slide contains animation. First click removes cover map, second click shows path actually taken and find location.

Lost Person Behavior　　　　　　　　　　　September 2014

Unit 7 – Subject Categories

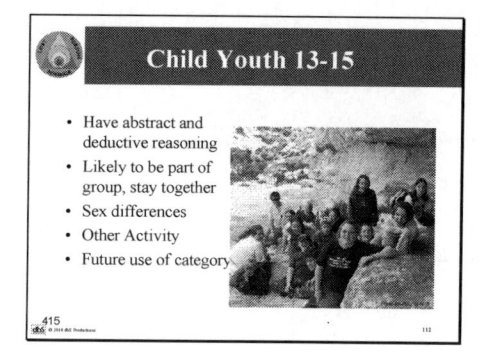

Slide 112

Child Youth 13-15
- Have abstract and deductive reasoning
- Likely to be part of group, stay together
- Sex differences
- Other Activity
- Future use of category

Text reference: 150-154. This subject category may not be used in the future. Some preliminary research shows youth hikers match adult hikers, and youth hunters match adult hunters. However, traits like staying together if in a group will remain important.

Slide 113

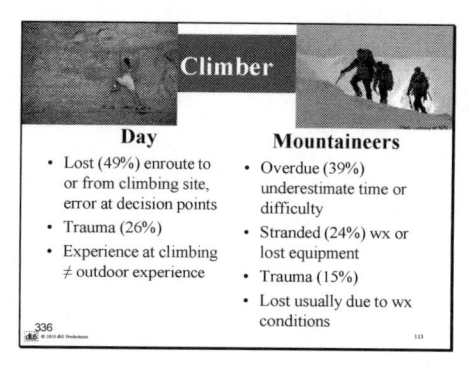

Climber

Day
- Lost (49%) enroute to or from climbing site, error at decision points
- Trauma (26%)
- Experience at climbing ≠ outdoor experience

Mountaineers
- Overdue (39%) underestimate time or difficulty
- Stranded (24%) wx or lost equipment
- Trauma (15%)
- Lost usually due to wx conditions

Text reference: 155-160

Slide 114

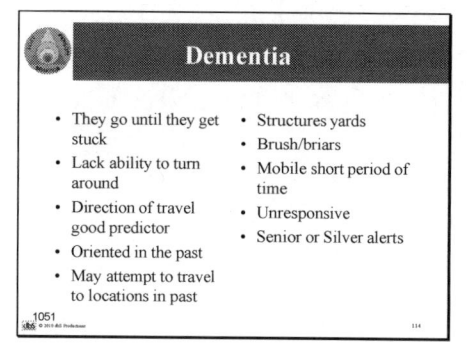

Dementia
- They go until they get stuck
- Lack ability to turn around
- Direction of travel good predictor
- Oriented in the past
- May attempt to travel to locations in past
- Structures yards
- Brush/briars
- Mobile short period of time
- Unresponsive
- Senior or Silver alerts

Text Reference: 161-169. Instructor's DVD contains 6 additional references.

Lost Person Behavior — September 2014

Unit 7 – Subject Categories

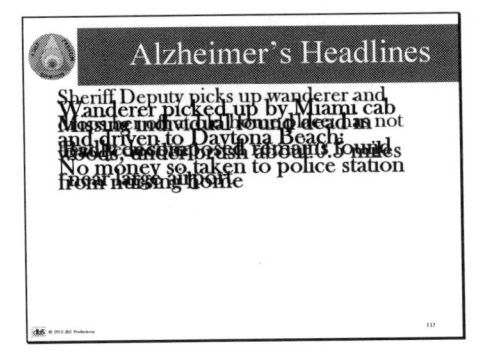

Slide 115

Animation, each click gives a different headline.

Slide 116

This may be the appropriate place to conduct the outdoor exercise if time permits. The exercise is fully described in Activity 7-10 Dementia field exercise in the Lost Person Behavior Instructor's Guide. The timing of the field exercise is not critical. So ideally it can be done just before a scheduled break.

The basis for "They go until they get stuck" is still unknown. But it makes sense considering several factors:

- Lack of short term memory in new location, means your reality is what you see. Makes turning around less likely.
- Virtual tunnel vision
- Known many with Dementia looking down at about 5-8 feet in front of them. Perhaps to look at the ground for safe walking since unable to commit obstacles to memory as much.

Lost Person Behavior   September 2014

Unit 7 – Subject Categories

Slide 117

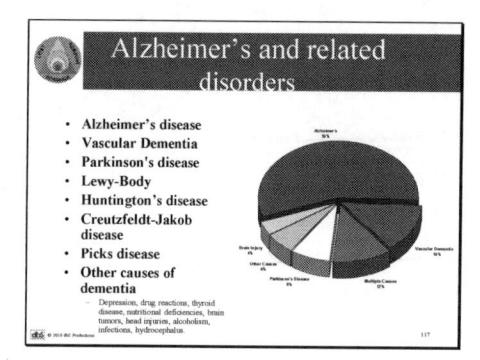

Not important to remember all of these disorders, but they all represent dementia. Everyone with Alzheimer's has dementia but not everyone with dementia has Alzheimers. In some cases caregiver may deny Alzheimer's if asked if the subject as another disorder. From a search and rescue perspective the wandering behavior is largely the same for all of the disorders.

Slide 118

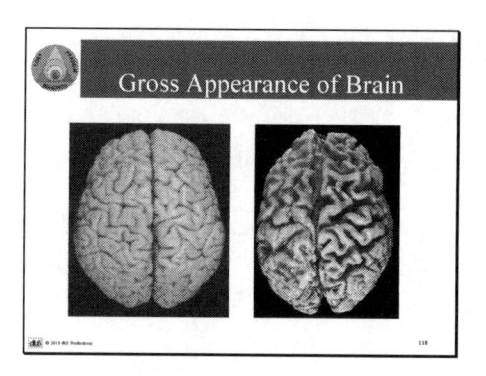

Animation: second click causes brain from someone with dementia to enter. First brain (left) is a normal brain. The link also provides access to the NIH video on dementia which explains more of the neurobiology.

Slide 119

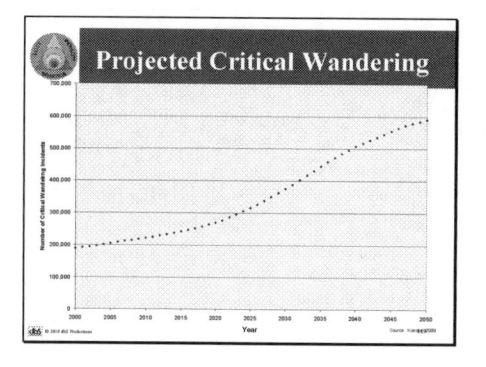

Projected curve of wandering based upon the projected increase of Alzheimer's in the United States. Largely based upon US Census projections of the increase in elderly Americans.

Unit 7 – Subject Categories

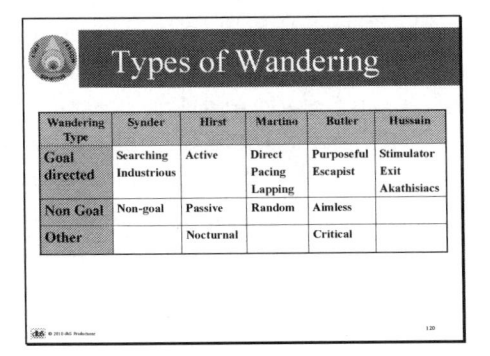

Slide 120

Types of Wandering

Wandering Type	Synder	Hirst	Martino	Butler	Hussain
Goal directed	Searching Industrious	Active	Direct Pacing Lapping	Purposeful Escapist	Stimulator Exit Akathisiacs
Non Goal	Non-goal	Passive	Random	Aimless	
Other		Nocturnal		Critical	

While many different terms exist to describe wandering they can be lumped into Goal and non-goal directed.

Slide 121

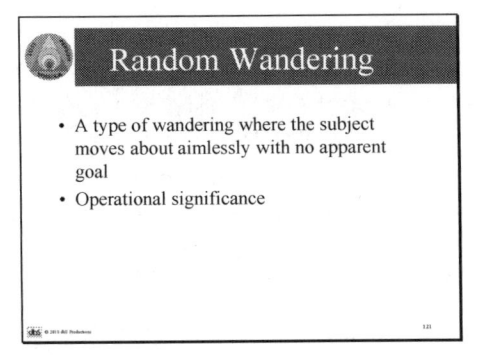

Random Wandering

- A type of wandering where the subject moves about aimlessly with no apparent goal
- Operational significance

Slide 122

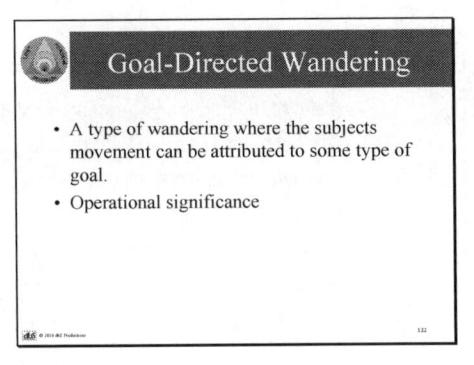

Goal-Directed Wandering

- A type of wandering where the subjects movement can be attributed to some type of goal.
- Operational significance

Lost Person Behavior

Unit 7 – Subject Categories

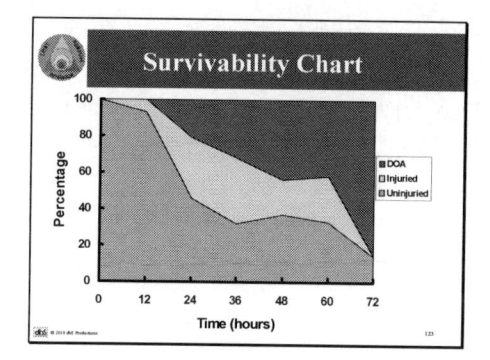

Slide 123

The graph presents a visual summation of nearly 100 lost Alzheimer's disease cases from Virginia data collected over several years. There is a narrow window where all subjects are located alive. Injuries and fatalities begin after 12 hours. The small percentage that remains uninjured even after 72 hours represents those cases that sought shelter in a structure (often abandoned) or were found hiding inside a residence. The chart is just a guide: inclement weather, clothing, and entry into water will greatly modify actual survival.

Slide 124

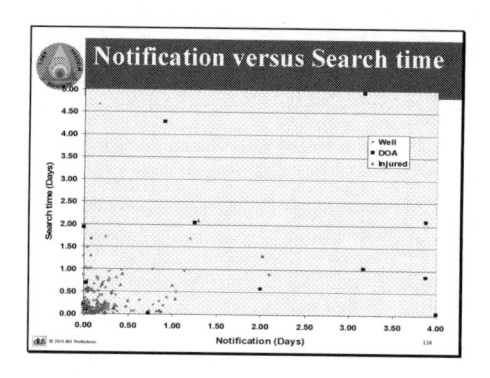

From ISRID data. Key point is when notified quickly, usually a quick find. Fatalities were associated with long delays in getting search started. Early fatalities were associated with drownings.

Slide 125

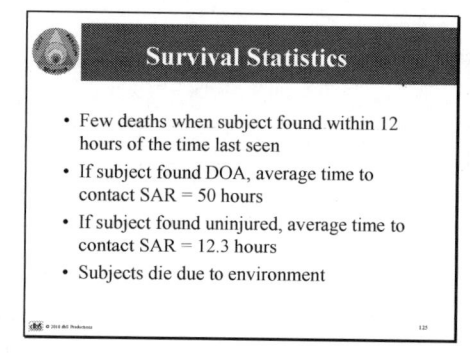

Survival Statistics
- Few deaths when subject found within 12 hours of the time last seen
- If subject found DOA, average time to contact SAR = 50 hours
- If subject found uninjured, average time to contact SAR = 12.3 hours
- Subjects die due to environment

Time to contact SAR involves three important components. The first is the amount of time it takes the family or caregiver to notice the patient is missing. The second component is the time it takes the caregiver to contact local law enforcement. The Alzheimer's Association's Safe Return program recommends that caregivers spend 15 minutes looking before contacting law enforcement. However, one study found 50% wait much longer. The final component is the time law enforcement take to initiate a search with trained SAR resources.

Unit 7 – Subject Categories

Slide 126

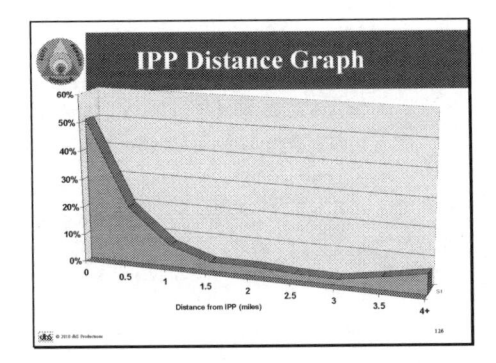

Presentation of the distance from the IPP data in a graphical format. Graph shows 632 cases and combines temperate, dry, and urban. Once again the data shows the high concentration of cases found within one mile. This data also agrees with a study that looked at cases reported by caregivers. They also found 80% of lost subjects within two miles.

Slide 127

Dementia

miles	Temperate		Dry	
	Flat	Mountains	Flat	Mountains
Count	175	95	15	14
25%	0.2	0.2	0.3	0.6
Median	0.6	0.5	1.0	1.2
75%	1.5	1.2	2.2	1.9
Max Zone	7.9	5.1	7.3	3.8

Review of statistical data for dementia in miles. In many cases instructors often present only the dementia presentation, so the full data is shown. If how to use the book is already shown, the instructor may elect to hide these slides.

Lost Person Behavior

Unit 7 – Subject Categories

Slide 128

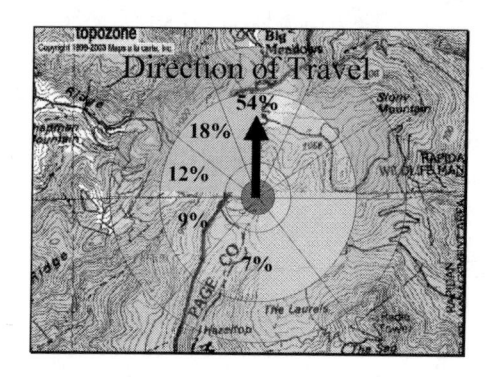

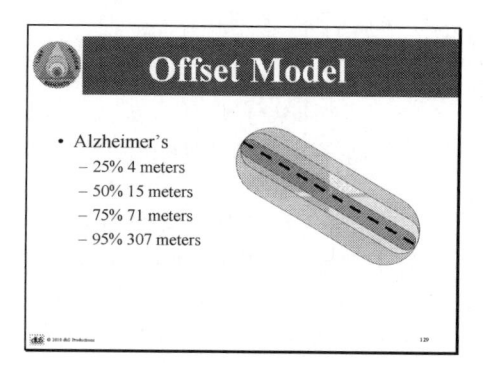

Slide 129

Offset Model

- Alzheimer's
  - 25% 4 meters
  - 50% 15 meters
  - 75% 71 meters
  - 95% 307 meters

Many of those subjects found off roads or trails (travel-aid) are located only a short distance off the travel-aid. The following table only looked at those subjects that were found at least a yard off of a travel aid. 50% of the cases are found within 15 meters of the travel-aid.

Slide 130

Environment of Find (Wild)

Bushes/Briars	29%
Creeks/Drainages	18%
Open Field	18%
House	18%
Road	7%
Woods	7%
Swamp	4%

The general pattern is "they go until they get stuck." Alzheimer's patients are one of the more demanding behavioral profiles due to their tendency to go deep into thick brush or briars. Finding subjects in creeks and drainages represents a path of least resistance until the natural topology catches the subject. Subjects found in open fields were typically found along fence lines. This data is somewhat different than that found in the book for find locations. However, it is helpful to see the large percentage found in bushes/briars.

Unit 7 – Subject Categories

Slide 131

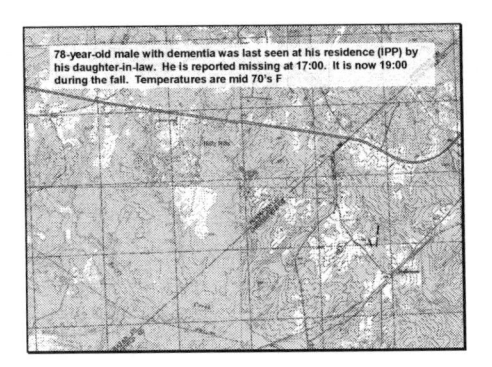

Next slide gives a tighter zoom and should be used for the actual map problem.

Slide 132

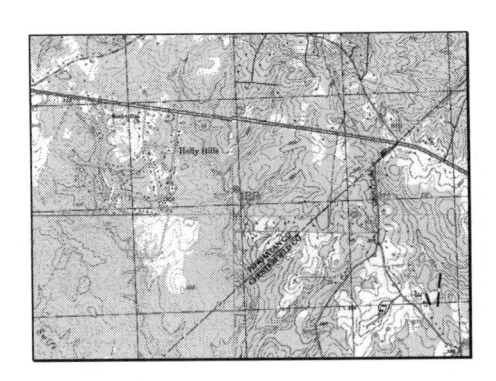

Slide contains animation. First click removes cover map. Second click reveals first find (summer) found by creek, third click (Fall) reveals find location of second search (few meters off driveway), following spring third search found at intersection of creek and route 60, then son builds fence with the gate facing the Northeast, next search found in heavy vegetation predicted by gate, fifth search gate but no latch (same general location), sixth search latch but no lock. The last find location is blue. This represents the search that triggered a request for a state mutual aid response. The local law enforcement agency sent an initial search team to the correct area. However, the search took place at night and they missed the subject. At that time they made the request for additional resources. While resources were mobilizing they search the area a second time and made the find.

Unit 7 – Subject Categories

Slide 133

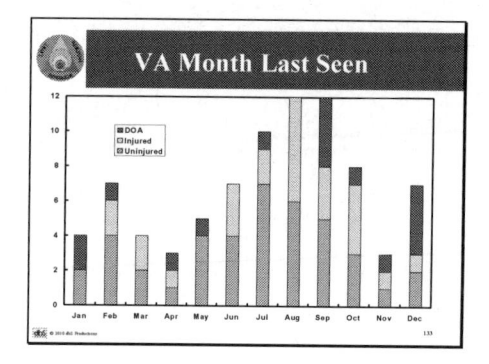

All data from Virginia only. Shows seasonal variation in incidents. Higher incidents in December often due to family stress during the holidays and February incidents occurred when temperature went above 70 degrees F.

Slide 134

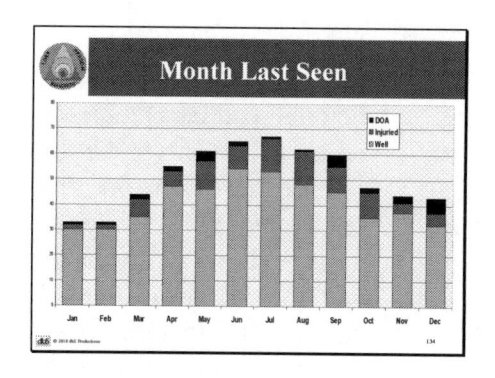

Data only from Northern Hemisphere, otherwise entire ISRID database for dementia searches (n=614)

Slide 135

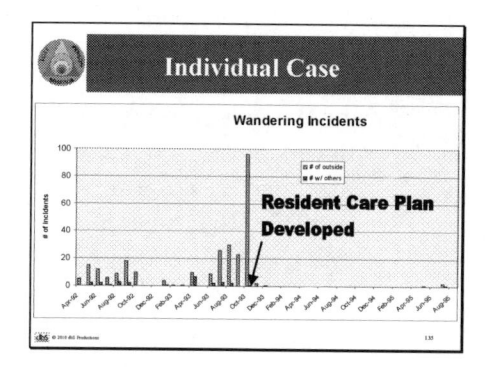

The following slide shows the number of incidents from a single person over the course of five years. The strong tendency to wander during the summer is clearly seen. The subject has been placed in a nursing home due to her wandering and dementia. Her previous employment prior to dementia was patient transportation in a hospital. It was her job to push wheelchairs from one location to the next. The blue bars represent where she pushed other nursing home residents out the door. During October of 1993 she wandered over 90 times. At that point the nursing home developed a non-restraint resident care plan, which worked well. Not until the summer of 1995 did her wandering pick

Lost Person Behavior — September 2014

Unit 7 – Subject Categories

back up some.

Slide 136

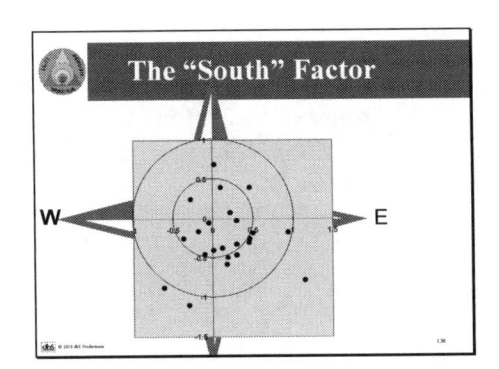

The "South" Factor

The slide shows incidents from Virginia. The center of the slide represents the IPP. The dots represent find locations plotted by direction. It can be observed that all but five of the incidents the subject was found more south than north of the IPP. This is most likely due to exit doors. Most of the incidents came from rural Virginia where it is the practice to build the front porch facing South. Therefore, the front door typically faces South. Support for this comes from three of the five incidents was found towards the north. In those three incidents the location of the exit door was known to be towards the north, and all of those subjects traveled more north than south. Other reasonable hypothesis would include be attracted towards light.

Unit 7 – Subject Categories

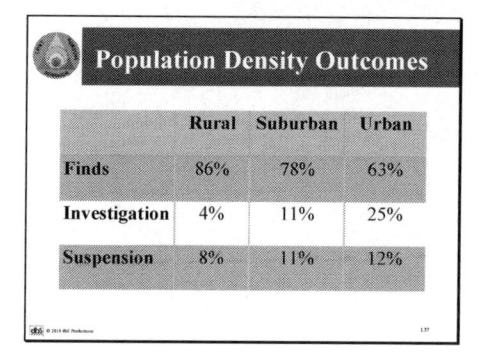

Slide 137

Population Density Outcomes

	Rural	Suburban	Urban
Finds	86%	78%	63%
Investigation	4%	11%	25%
Suspension	8%	11%	12%

Differences between rural, suburban, and urban searches were examined. No differences were seen for distance traveled from the PLS, survivability, severity, or find location. The only difference was between the types of finds in rural versus urban cases. In urban cases more finds occur through law enforcement investigation.

Slide 138

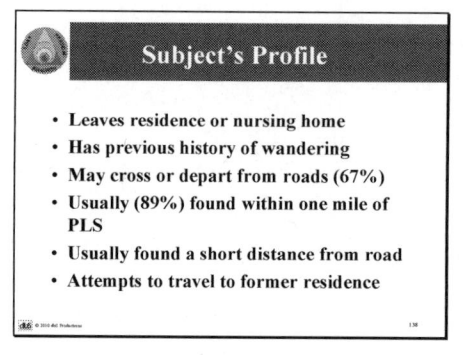

Subject's Profile

- Leaves residence or nursing home
- Has previous history of wandering
- May cross or depart from roads (67%)
- Usually (89%) found within one mile of PLS
- Usually found a short distance from road
- Attempts to travel to former residence

The above represents the general behavioral profile similar to a FBI behavioral profile of criminal conduct. It was generated in a similar manner.

Slide 139

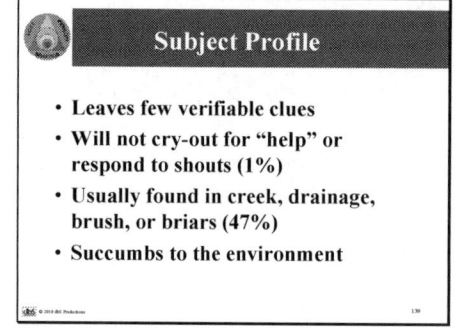

Subject Profile

- Leaves few verifiable clues
- Will not cry-out for "help" or respond to shouts (1%)
- Usually found in creek, drainage, brush, or briars (47%)
- Succumbs to the environment

The above represents the general behavioral profile similar to a FBI behavioral profile of criminal conduct. It was generated in a similar manner.

Unit 7 – Subject Categories

Slide 140

Shift Supervisor's Initial Action
- Receive initial report from initial officer
- Implement initial command structure
- Establish location of search base or staging area for incoming resources
- Increase investigation
- Determine search urgency

Instructor may pass out Alzheimer's Association Guide for Law Enforcement handout. The Alzheimer's Association's Safe Return may provide several different types of assistance to the law enforcement official. If the lost person is registered with the Safe Return, the program they will provide a photograph, subject description and medical description by fax to law enforcement. Safe Return can also provide a representative from one of the local Alzheimer's Association chapters to counsel the lost patient's family. (In some parts of the country, Safe Return can fax bulletins to a much broader network of hospitals, shelters and other community agencies, plus media outlets.)Subject must be reported to both surrounding jurisdictions and entered into the NCIC network. Entering the subject into NCIC is especially important for mild Alzheimer's and urban cases where the lost subject may have utilized transportation to move outside of the search area. A copy of the NCIC report form is included in the resource kit as a possible handout.

Slide 141

Supervisor's Notifications
- Vary depending upon departmental SOP
- Contact Alzheimer's Association Safe Return Program (800) 625-3780
- Local hospitals, shelters, morgues, medical examiners office.
- Radio report to surrounding jurisdictions, NCIC network

Instructor may pass out Alzheimer's Association Guide for Law Enforcement handout. The Alzheimer's Association's Safe Return may provide several different types of assistance to the law enforcement official. If the lost person is registered with the Safe Return, the program they will provide a photograph, subject description and medical description by fax to law enforcement. Safe Return can also provide a representative from one of the local Alzheimer's Association chapters to counsel the lost patient's family. (In some parts of the country, Safe Return can fax bulletins to a much broader network of hospitals, shelters and other community agencies, plus media outlets.)Subject must be reported to both surrounding jurisdictions and entered into the NCIC network. Entering the subject into NCIC is especially important for mild Alzheimer's and urban cases where the lost subject may have utilized transportation to move outside of the search area. A copy of the NCIC report form is included in the resource kit as a possible handout.

Slide 142

Contact Trained SAR Resources

- Bloodhounds
- Air-scent dogs
- Helicopters
- Fixed wings
- Search management
- Man-trackers
- Field team leaders
- Field team members
- Mounted SAR teams
- Bike teams
- other specialized SAR resources

Trained SAR resources allow detection of clues, searching at night, and may bring years of SAR experience to the search. However, most resources are volunteers and may require 1-3 hours to mobilize and travel to the search site. Therefore, resources must be contacted. Requesting trained SAR resources significantly reduces liability (most lawsuits involve an insufficient response) to law enforcement.

Slide 143

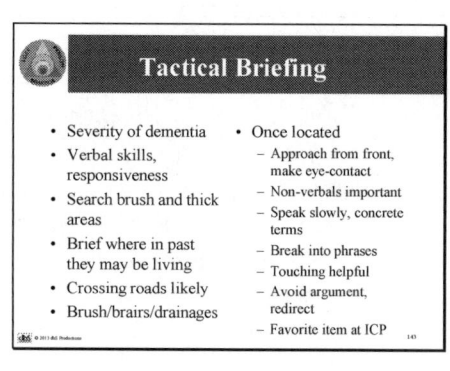

Tactical Briefing

- Severity of dementia
- Verbal skills, responsiveness
- Search brush and thick areas
- Brief where in past they may be living
- Crossing roads likely
- Brush/brairs/drainages
- Once located
  - Approach from front, make eye-contact
  - Non-verbals important
  - Speak slowly, concrete terms
  - Break into phrases
  - Touching helpful
  - Avoid argument, redirect
  - Favorite item at ICP

Unit 7 – Subject Categories

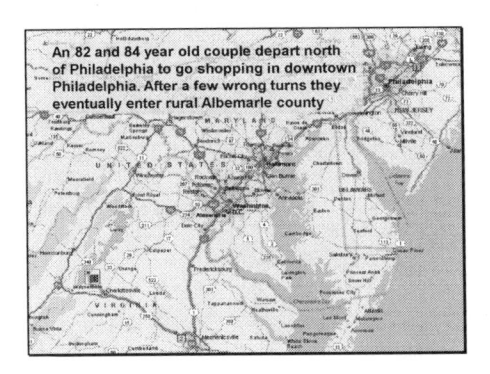

Slide 144

Full description of incident is in the Instructor's Guide Activity 7-7

An 82 and 84 year old couple depart north of Philadelphia to go shopping in downtown Philadelphia. After a few wrong turns they eventually enter rural Albemarle county

Slide 145

They headed north up county road 810, but missed the sharp right the road makes.

Google Image of Browns Cove. The yellow line is route 810 which they would have traveled heading North. Note the sharp turn to the right. They did not make this

Slide 146

They continued going straight, past an old estate and into a road going back to a cabin. They parked at the cabin (end of road).

The couple continued straight leaving 810 (at the sharp bend) and going straight onto the Brown Gap Turnpike. The actual turnpike turns off to the right (first right) and heads up the mountain. The couple continued going straight, passed a large house by the pond, entered the now gravel drive into the woods and parked in the driveway of a cabin where the road ended (yellow dot).

Lost Person Behavior

Unit 7 – Subject Categories

Slide 147

Slide 148

Closer view of the IPP. The couple had pulled in on Thursday night and parked. The car was parked in the driveway of a weekend home (able to see the green roof). The owner of the cabin arrived late Friday afternoon and found the car. He immediately called police. Running the plates found a match for the missing couple in NCIC (National Crime Information Center). The search started Friday night with scent-discriminating dogs. That only provided some limited activity around the cabin itself. At first light on Saturday the formal search started. The road the couple used to get to the cabin is indicated by the yellow lines. The white line shows a road that existed to the top of the mountain that did not appear on the topographic map. The dirt road seen to the left of that road is from logging activity that occurred after the search. It did not exist at the time. The thinner white line running horizontally is the unimproved road that can be seen on the topographic map.

Lost Person Behavior

Unit 7 – Subject Categories

Slide 149

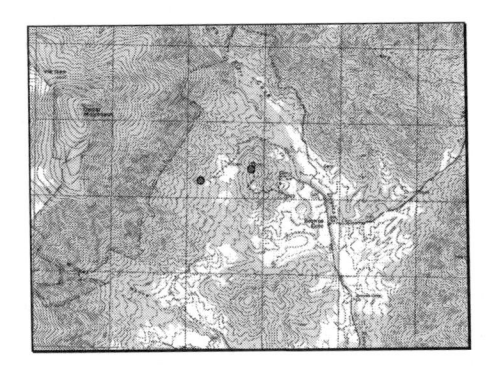

Topographic map of the same area. Finds are shown. Slide is animated to show finds on each click. It is suggested to use the highlighter pointer feature to demonstrate the reflex tasking.

Slide 150

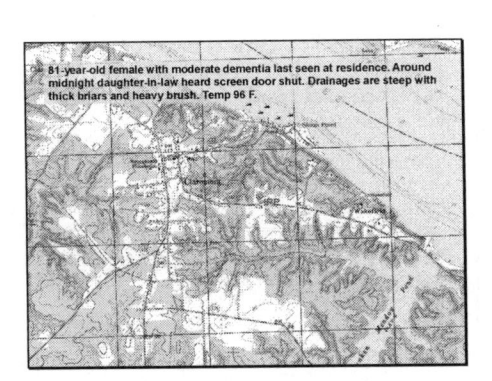

An 81-year-old female mother with moderate Alzheimer's disease is missing. She currently lives at her son's home where she has been staying for the past six years. At around midnight the daughter-in-law thought she heard the screen door slam shut. This door faces south. When she awoke in the morning her mother-in-law was missing. She is dressed only in her pajamas and did not bring her slippers so is believed to be barefoot. She also suffers from diabetes (non-insulin dependent) and poor eyesight. The drainages into the James River are extremely steep with thick briars and heavy brush. It is a hot summer's day with highs in the upper 90's with high humidity. 60% chance of late afternoon thunderstorms. She has not wandered before.

Lost Person Behavior · September 2014

Unit 7 – Subject Categories

Slide 151

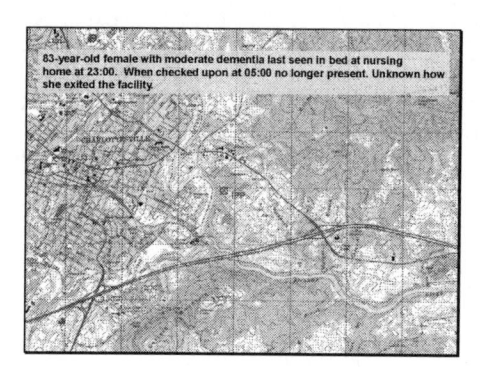

Two finds come from the same IPP (but different subjects). In the first, the subject returned to their neighborhood and in the second they were found in the closest drainage to the IPP. See IG 7-9 for exact locations.

Slide 152

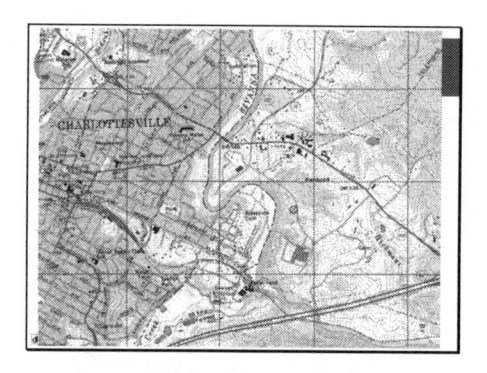

Slide 153

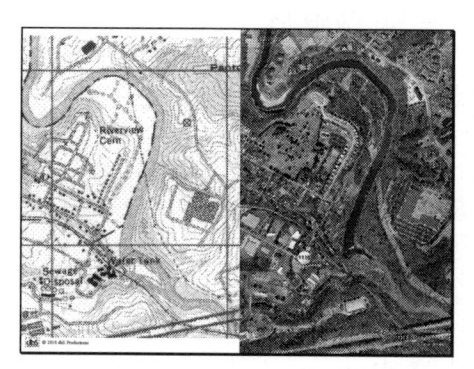

Unit 7 – Subject Categories

Slide 154

Despondents

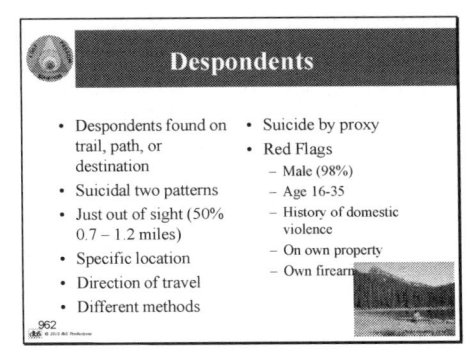

- Despondents found on trail, path, or destination
- Suicidal two patterns
- Just out of sight (50% 0.7 – 1.2 miles)
- Specific location
- Direction of travel
- Different methods
- Suicide by proxy
- Red Flags
  - Male (98%)
  - Age 16-35
  - History of domestic violence
  - On own property
  - Own firearm

Text Reference: Page 170-177. Additional resource references in Instructor's DVD.

Slide 155

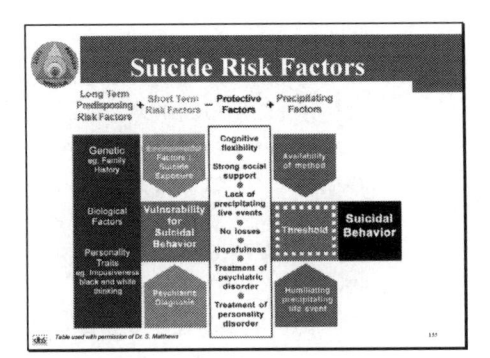

Suicide Risk Factors

A basic description of the graphic is provided in the textbook on page 170. The purpose of the graph is to provide some additional insight on the likelihood the subject is actually attempting a suicide. The graphic was used with permission from Dr. S. Matthews and the pamphlet the graphic was derived from can be found in the Instructor's Resource DVD in the Resource Books and Reports directory, then the Despondent directory, finally the PDF is called SuicideRiskBooklet.PDF

Slide 156

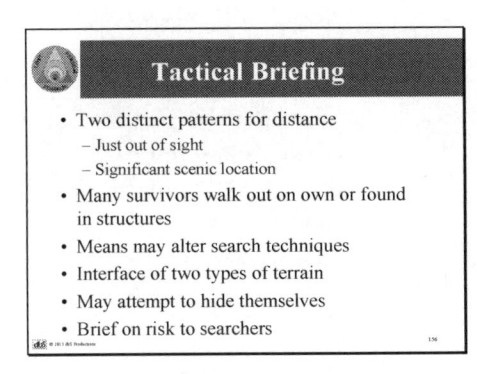

Tactical Briefing

- Two distinct patterns for distance
  - Just out of sight
  - Significant scenic location
- Many survivors walk out on own or found in structures
- Means may alter search techniques
- Interface of two types of terrain
- May attempt to hide themselves
- Brief on risk to searchers

Lost Person Behavior

Unit 7 – Subject Categories

Slide 157

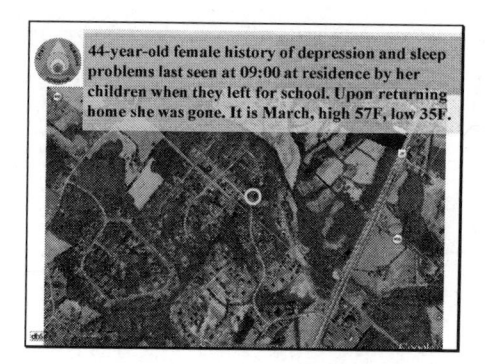

Google Earth view of the search area and the IPP (subject's residence). Note the development does not appear on the topographic maps. Instructor's Guide Activity 7-11

Slide 158

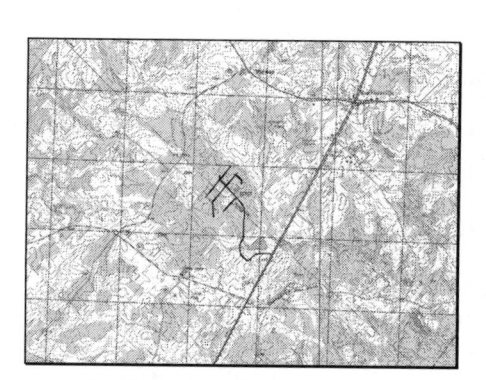

A 44-year-old female was last seen by her children at 09:00 in the morning when they left for school. She has a history of depression and sleep problems. The depression and sleep problems have gotten worse over the last two weeks. She normally did not prepare breakfast for the children but this morning she did. The children reported she was more affectionate than normal. When they returned home, she was not present. Her husband is out of town on a business trip, this was confirmed as a legitimate business trip. No suicide note was left. However, the previous day she called two of her close friends and expressed her appreciation for their friendship.

The children reported that she was wearing beige fleece jacket, black running pants, black sneakers, and a black sweat band.

Missing from the house is a picture of the family and the husband's handgun. She has her own handgun and from time to time shots on a range. The missing handgun fires a .357 Magnum ammunition. While she had been car camping with the family she did not particularly like the woods or bugs. The family was originally from New York and they ran a popular New York Pizzeria in town. During the day she was often seen walking along the roads in the subdivision. No marital or financial difficulties. Investigation shows no activity on cell phone or credit cards.

Slide 159

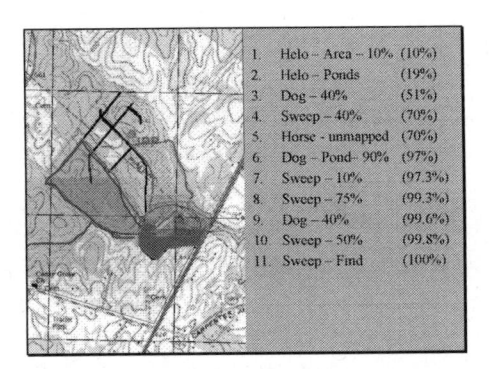

Slide is animated. Each click shows another task (described in the list) that had a chance of locating the subject. This search was unusual, but it does show that even with a PODcum of 99.8% the chance remains (and did) of still finding the subject.

Slide 160

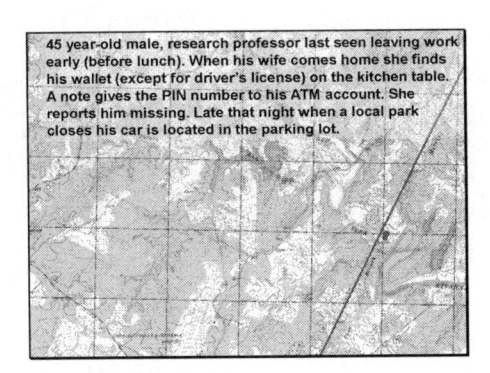

A 45-year-old research professor in neurobiology employed by the local university was last seen leaving work early (before lunch). When his wife comes home she finds his wallet (except for driver's license) on the kitchen table along with a short note that only gives the PIN number to his ATM account. She reports him missing. Late that night when the local park closes a car is still in the parking lot. The plates are run and it is determined to be his car.

This was the day he was expected to hear if his grant was going to be funded. If the grant was funded he was going to be offered a tenure track position. If the grant was not funded he was going to need to

Unit 7 – Subject Categories

find a new job at a different university. His grant was not funded.

His wife also holds a research position. Much of their lives revolve around research interests. She is happy with her position and likes the area. They have two children. They all have visited Chris Green Lake Park once before to go swimming. He did not hike any of the trails. His car is located in the first parking lot that has access to a trail. The trail is not located on the map and will need to be shown to the participants.

He did not suffer from depression but was known to be impulsive. He had been apprehensive about the grant. He had no other medical conditions. He did not own or have access to a firearm. Nothing of interest was found in the car.

Anyone familiar with typical neuroscience labs will know they often contain a wide array of highly toxic poisons used in research. These substances are not under any strict controls so determining what might be missing from a lab would be difficult. During the actual search this was pointed out to local law enforcement and the question was asked if they needed to bring in a state hazardous material team. Since all of the common neurotoxins found in a typical lab must be ingested they pose little harm if a responder wears gloves and does not taste any white powders.

Instructor's Activity Guide 7-12

Unit 7 – Subject Categories

Slide 161

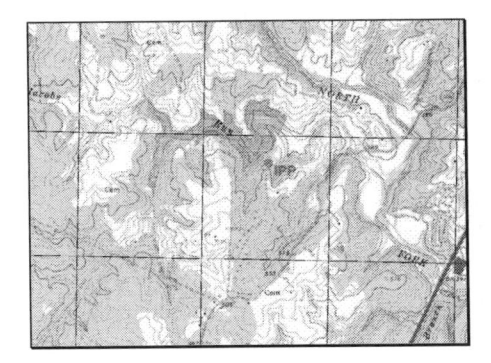

Topographic map of the area. Note minimal details. Ask class what do most local parks have? More detailed map is the reply you want.

Slide 162

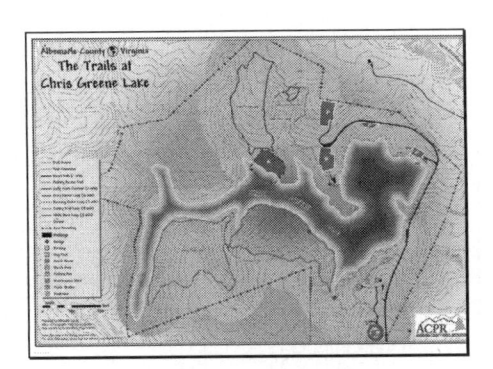

Trail map of the park.

Slide 163

Lost Person Behavior   September 2014

Unit 7 – Subject Categories

Slide 163

Slide 164

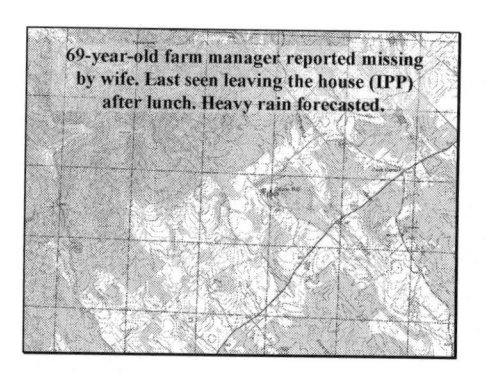

Overall view of the search area that matches printed map the best. Use this slide to present the scenario. This scenario is also excellent for investigation into factors that predict suicide. The what really happened also has some unique twists. See Instructor's Guide Activity 7-13

Slide 165

A 69-year-old farm manager is reported missing by his wife. He was last seen leaving the house (which is located on the farm property) after lunch. He did not return for dinner. After some quick searching on the part of friends he was reported missing.
He has a history of depression. The previous night he called his psychiatrist to schedule an appointment. The psychiatrist was not able to meet and called in a prescription for the subject. The subject never picked up the prescription. It is typical for him to get depressed on the anniversary of his brothers death (by suicide) which is today.
He has been the farm manager for the

Lost Person Behavior

Unit 7 – Subject Categories

Castle Hill property for 30 years and is well liked by the property owners, other farm workers, and everyone in the surround community. He has always done an excellent job and is considered somewhat of a perfectionist. He knows the property extremely well and it is considered impossible for him to become lost on the property. He does hunt and the wife cannot locate his hunting rifle.
All other investigation points towards him being in the field

Slide 166

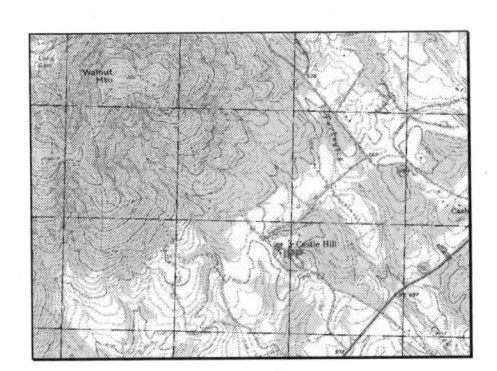

Use this slide if allowing students to determine tasking for map problem.

Slide 167

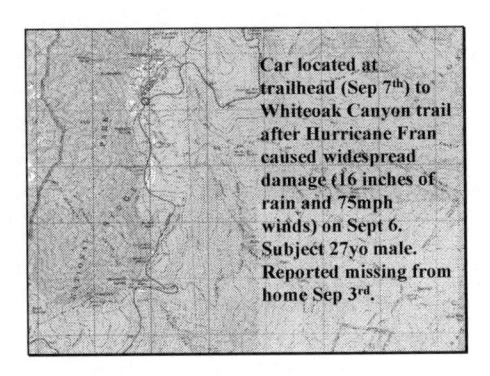

On September 6 Hurricane Fran struck central Virginia with 75 mph winds and 13 inches of rain. This caused significant destruction to trees at higher elevations, especially in Shenandoah National Park. A road cutting crew was able to reach the Whiteoak Canyon trailhead parking lot on September 7th. They found a red Hyundai with a notice to evacuate that was left prior to the storm on September 5th. Unfortunately, when they ran the plates and contacted the rental company on the 5th the rental company provided the wrong information regarding the renter. However, on the 7th the learned the car had been rented by a 27-year-old male who had been reported missing on

Unit 7 – Subject Categories

September the 3rd. In the car was new camping equipment, two boxes of ammunition, a box for a 9mm handgun, a sales receipt for all the gear dated August 30th. The handgun could not be located. A review of credit card charges showed the last activity was August 30 at the Big Meadows lodge for a dinner and paying for a room at Skyland the night of the 30th. A review of the parks videotape at the entrance stations showed his car entering on the 30th. Therefore, it was possible to talk to the entrance station ranger on duty at that time. The ranger was able to recall the subject and him asking where the best waterfalls were in the park. The ranger directed him to the Whiteoak Canyon trail. He was not familiar with the park.

The family reported he had received a set back to his career in the State Department but he was looking forward to his first assignment as a Foreign Service officer to Athens. He had mentioned doing some hiking and camping to help relax. Additional investigation with the State Department revealed that he had just failed a second polygraph test regarding mishandling of classified material while a naval officer in Athens. He had failed the first polygraph test when applying to work for the CIA in 1995. The CIA notified the FBI, but the FBI had only notified the State Department a month prior to his scheduled deployment to Greece. Upon learning this the State Department administered a second polygraph test which he failed. He then lost access to all classified material and restricted areas.

The search started on September the 7th. The combination of previous significant rains, the 13 inches from Hurricane Fran, and the high winds caused many of the hardwood trees to be toppled over with full foliage. This created an almost impenetrable tangle of fallen trees on the roads and trails throughout the park. Most of the park was closed and access to

Unit 7 – Subject Categories

buildings was limited. Electricity was provided by generators. Students when deploying resources will most likely want to deploy a team down into Whiteoak canyon. This will be difficult. Instructor's Guide Activity 7-14

Slide 168

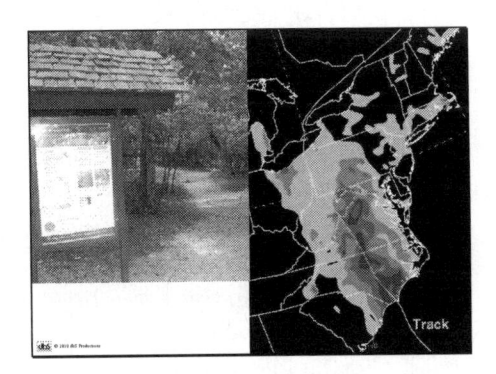

Picture of trailhead and illustration of rainfall from hurricane. The blue dot covers the location of the search and the purple ring is the location of the park,

Slide 169

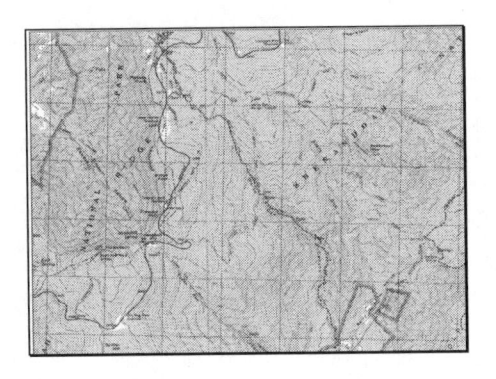

The information from the family pointed more to a lost hiker scenario. However, the purchase of the gun quickly pointed the search towards the despondent scenario. Since a federal law enforcement agency was conducting the search and investigation, the State Department was forthcoming with information. Key facts and factors were known to search planners early in the search.

While search strategy using lost person behavior profiles was straight forward, carrying out the strategy due to the Hurricane was highly problematic. The need to search down Whiteoak Canyon was clear along with special emphasis around the waterfalls. During the actual

Unit 7 – Subject Categories

search the first team simply gave up after spending hours to travel about a half mile (one kilometer). The second team was an air-scent dog team and the handler broke her leg. After that no further teams were dispatch until a maintenance crew spent two days with chainsaws to clear the trail. Access from the bottom of the trail also failed. Once the trail was cleared teams could move down the trail. No air-scent dog alerts occurred. Teams were dispatched to search around each of the waterfalls but after hours in the area they reported PODs of 5-10%. A Blackhawk helicopter was also used to fly at tree top level. The search was active from September 7th-15th and then suspended. On October 28th a park visitor stepping off the trail to relieve himself spotted yellow highlighters that belonged to the subject and then spotted the deceased subject. He was 40 yards off the trail in a rocky location. By this time the leaves had started to drop off the fallen trees and visibility was increasing.

Slide 170

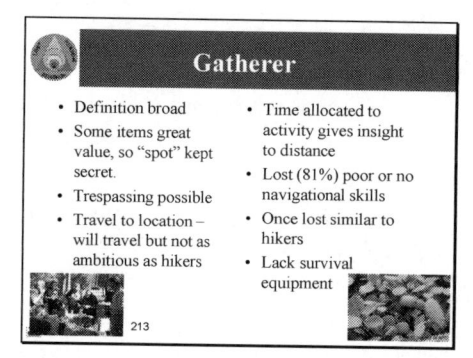

Gatherer
- Definition broad
- Some items great value, so "spot" kept secret.
- Trespassing possible
- Travel to location – will travel but not as ambitious as hikers
- Time allocated to activity gives insight to distance
- Lost (81%) poor or no navigational skills
- Once lost similar to hikers
- Lack survival equipment

Text reference: Page 178-182

Lost Person Behavior — September 2014

Unit 7 – Subject Categories

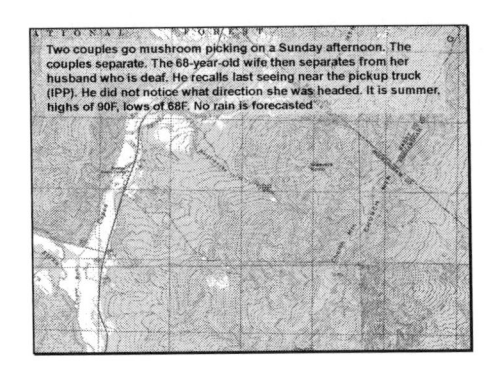

Slide 171

Two couples go mushroom picking on a Sunday afternoon. The couples separate. The 68-year-old wife then separates from her husband who is deaf. He recalls last seeing near the pickup truck (IPP). He did not notice what direction she was headed. It is summer, highs of 90F, lows of 68F. No rain is forecasted.

Two elderly couples go out Sunday afternoon to go mushroom picking. Each of the couples separate. Eventually, the 68 year old wife also separates from her husband who is deaf. The husband returns to the pick-up truck which is the last location he clearly remembers seeing her (IPP). They were never that far from the pickup truck. The husband tried shouting but without any luck. He did not notice what direction she was last headed. Additional investigative information is rather limited. Nothing suggests that either of them had dementia. She did not have any other health issues. The couple was relatively new at mushroom gathering with the other couple having more experience. It was summertime with highs in the low 90s F and late afternoon thunderstorms. Although, the search area did not receive any rain. She only has a small basket and shovel with her. They left the water and food in the pickup truck. She is not familiar with this location.

This search should be easy for the students to solve if they correctly use reflex tasking concepts. However, at the time (1980) none of these concepts had been developed. In fact, the search required 7 days and over 800 people (daily count x days). At the time the practice was to conduct grid searches starting at the IPP and expanding out. See Instructor's Guide 7-15.

Lost Person Behavior — September 2014

Unit 7 – Subject Categories

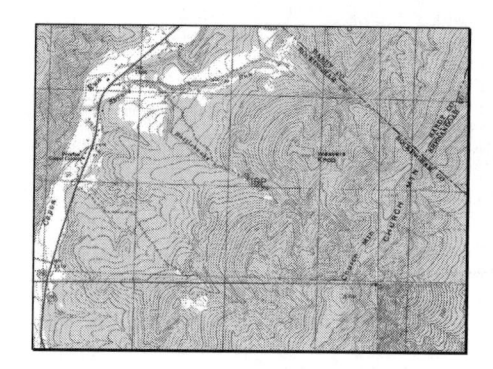

Slide 172

Slide 173

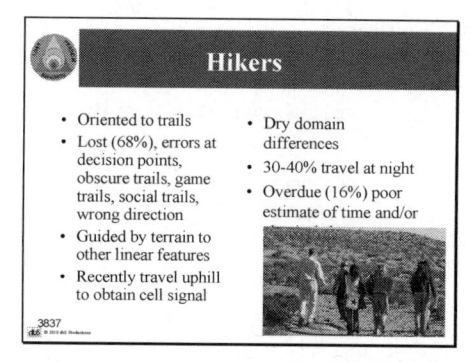

Text reference: 183-188

Hikers
- Oriented to trails
- Lost (68%), errors at decision points, obscure trails, game trails, social trails, wrong direction
- Guided by terrain to other linear features
- Recently travel uphill to obtain cell signal
- Dry domain differences
- 30-40% travel at night
- Overdue (16%) poor estimate of time and/or

Slide 174

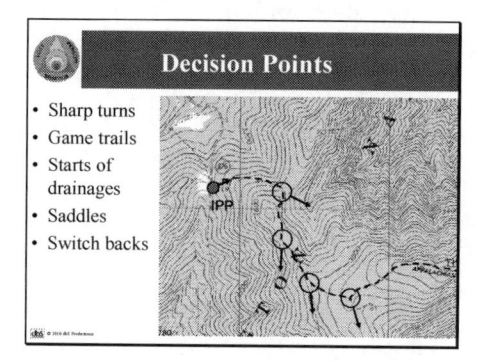

Decision Points
- Sharp turns
- Game trails
- Starts of drainages
- Saddles
- Switch backs

This slide and the next one are repeated from unit three. The slide is repeated here in case an abbreviated course is being given focused on only the subject categories. The concept is critical to understanding lost hikers.

Lost Person Behavior

Unit 7 – Subject Categories

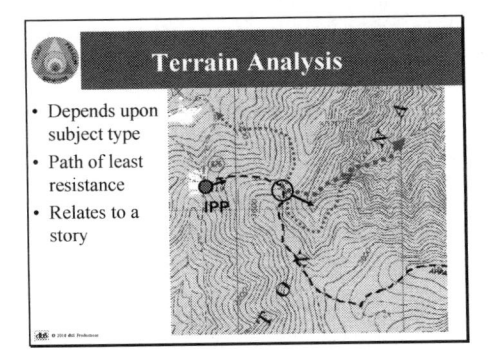

Slide 175

Terrain Analysis
- Depends upon subject type
- Path of least resistance
- Relates to a story

Repeated slide. See unit three for full details.

Slide 176

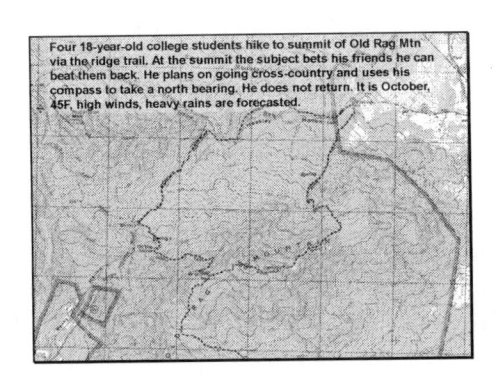

Four 18-year-old college students decided to hike to the summit of Old Rag Mountain and then turn around and return to the trailhead parking lot. However, once at the summit the subject bets his friends he could beat them back to the car. The friends returned via the ridge trail three hours later to the parking lot. The subject took out his compass set it on North and planned to go cross-country to the Weakley Hollow road and then return to the parking lot. He was last seen on the summit getting ready to head North at noon. It is October, 45 F, high winds, and heavy rains are forecasted.

The subject was a healthy 18-year-old but had suffered a dozen broken bones over his childhood. None of these fractures were from pathological factors but instead from daredevil activities such as bicycle riding, skateboarding, climbing trees, etc. The subject is dressed in jogging shorts, a polo shirt, and a sweater. He is wearing old tennis shoes. Other than a compass he has no other gear with him. He has been on previous hikes but no survival or navigation training or experience.

The summit of Old Rag Mountain is filled with large boulders, ledges, cliff faces, and mountain laurel thickets. Soon after the search started a 1,000 foot ceiling developed around the 3268 foot mountain.

Lost Person Behavior   September 2014

Unit 7 – Subject Categories

Visibility dropped to about 40-feet. A picture is included that shows the summit along with an aerial overview.

Slide 177

View near the summit of Old Rag along the ridge trail.

Slide 178

Old Rag MTN view looking south. This shows the terrain the subject would have to navigate if going due North.

Lost Person Behavior   September 2014

Unit 7 – Subject Categories

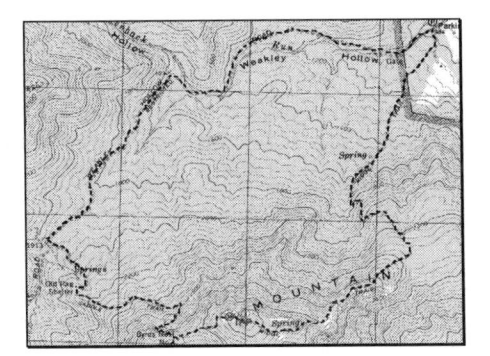

Slide 179 — Map that should be used for discussion.

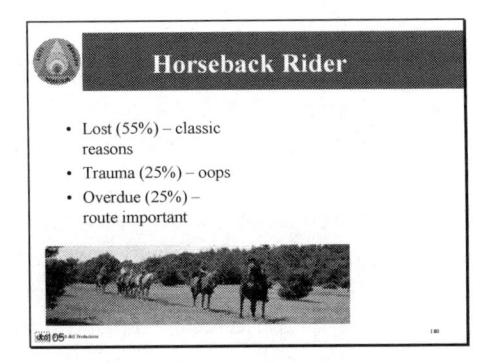

Slide 180 — Text reference: 189-192

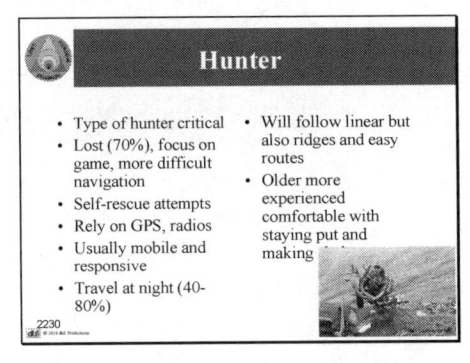

Horseback Rider
- Lost (55%) – classic reasons
- Trauma (25%) – oops
- Overdue (25%) – route important

Slide 181 — Text Reference: 193-198. Since hunting practices vary widely from region to region, this section should be customized to best match the hunting practices found in your region.

Hunter
- Type of hunter critical
- Lost (70%), focus on game, more difficult navigation
- Self-rescue attempts
- Rely on GPS, radios
- Usually mobile and responsive
- Travel at night (40-80%)
- Will follow linear but also ridges and easy routes
- Older more experienced comfortable with staying put and making

Lost Person Behavior — September 2014

Unit 7 – Subject Categories

Slide 182

A party of six is bear hunting. The group breaks up into pairs. A 52-year-old male traveling uphill starts to feel winded and tells his partner he needs to rest. He was last seen at 09:00 sitting against a tree. The partner cannot precisely locate where he last saw the subject. It is somewhere within the black circle.

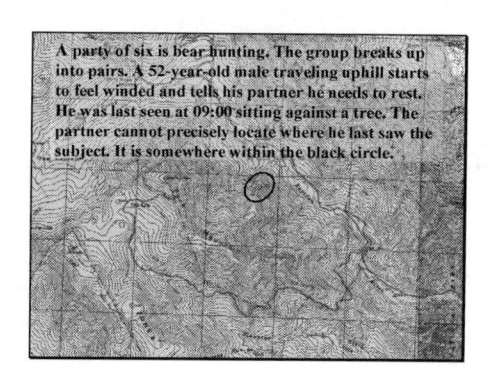

This particular search is somewhat unique since it reflects the medical scenario instead of a lost scenario. It is also somewhat unique in the fact the IPP was not precisely known. Participants who have mastered reflex tasking should have no problem making a fast find. For that very reason this problem might be better suited for the presentation technique of passing out the map and letting students work on it in small groups or individually. All students should be able to locate the subject and it will be instructive to see overall strategies.

A party of six is bear hunting. The group breaks up into pairs. The plan is to drive any bears to a waiting hunter. A 52-year-old male traveling uphill starts to feel a little winded and tells his hunting partner that he will rest for just a little bit and then catch back up. He was last seen at 09:00. They never meet back up and he does not return to the cars that are parked along the Rapidan Road. The hunting partner is not able to determine exactly on the map where he last saw the subject but can indicate a general area. The group spent the day much higher on the mountain and then returned to their cars at 16:30. It is December with highs of 45F and low of 25F. No snow is on the ground.

The subject has no physical ailments and is in average shape, which in not adequate for climbing the mountains. He is well dressed for hunting. He is wearing a flannel shirt, sweater, insulated overalls, and a jacket. He is wearing a hat and gloves. He is carrying a rifle, a handful of shells, a bottle of water, some snacks, a package of cigarettes, and a lighter. He has high blood pressure and high cholesterol for which he is taking medication. He has hunted in this area before and knew that all he needed to do was travel downhill to the road. He does not have a map or compass. He does not have a cell phone, nor is any coverage available.

Unit 7 – Subject Categories

The student map on the PowerPoint slide simply indicates the general area his hunting partner pointed out as being the possible IPP. Which he described as "walking up the ridge" and staying out of the drainage. He is willing to try to locate the IPP again on foot, but not sure if he can do it. The hunting party did fire shots, but got no response. The hunting party was not likely to hunt in the National Park and would stay in the Wildlife Management area.
See Instructor's Guide Activity 7-17.

Slide 183

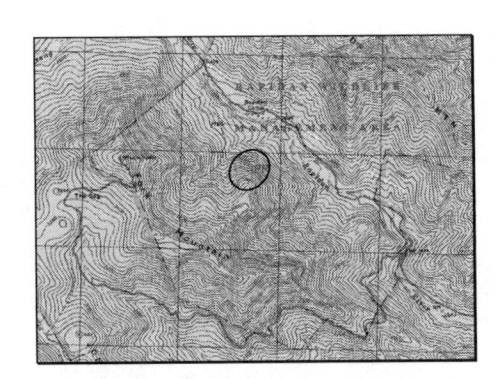

Closer view showing the general area of the IPP. The red line is the National Park boundary. The hunters are not expected to cross this boundry.

Slide 184

Medical "Scenario"
- Normal cognitive ability
- Scenario based
- Hikers, hunters, snow-mobiliers, bikers
- Cardiac (35%), heat (16%), hypothermia (13%), diabetic (10%), medications (6%)
- 75% within 1 km
- Linear features (53%), structure (12%), water (12%)

Lost Person Behavior — September 2014

Unit 7 – Subject Categories

Slide 185

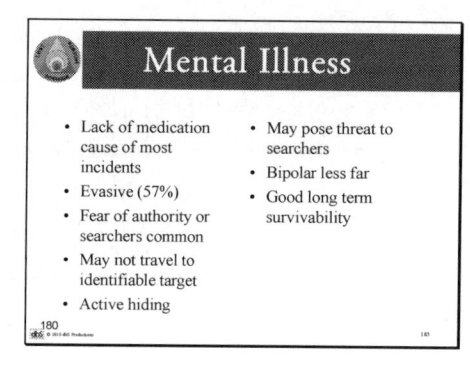

Mental Illness

- Lack of medication cause of most incidents
- Evasive (57%)
- Fear of authority or searchers common
- May not travel to identifiable target
- Active hiding
- May pose threat to searchers
- Bipolar less far
- Good long term survivability

Text reference: 199-205

Slide 186

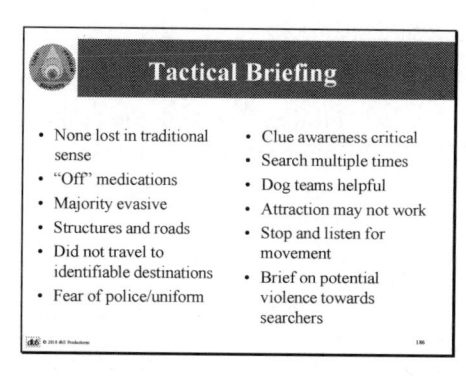

Tactical Briefing

- None lost in traditional sense
- "Off" medications
- Majority evasive
- Structures and roads
- Did not travel to identifiable destinations
- Fear of police/uniform
- Clue awareness critical
- Search multiple times
- Dog teams helpful
- Attraction may not work
- Stop and listen for movement
- Brief on potential violence towards searchers

Slide 187

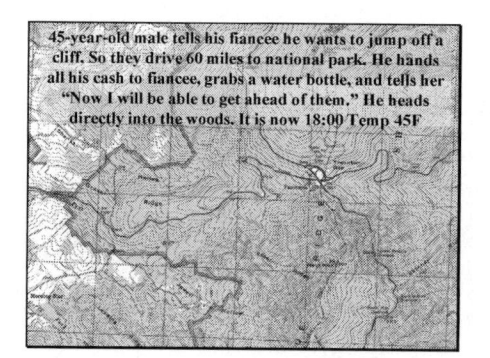

45-year-old male tells his fiancée he wants to jump off a cliff. So they drive 60 miles to national park. He hands all his cash to fiancée, grabs a water bottle, and tells her "Now I will be able to get ahead of them." He heads directly into the woods. It is now 18:00 Temp 45F

A 45-year-old male with a history of mental illness has been progressively getting worse. He has reported auditory hallucinations to his fiancée. It appears he has stopped taking his medications. The family does not approve of the fiancée and this is an additional source of stress. The previous weekend he attempted to drive to a national park in order to find a cliff to jump off. So he and his fiancée set out to find a national park. Departing from Fredericksburg, VA (50 miles away from the IPP) they could not find the park due to foggy conditions, no planning, and no map. A week later on Sunday he and his fiancée once again set out for the national park. They park the car inside the park in the

Lost Person Behavior   192   September 2014

Unit 7 – Subject Categories

parking lot of a defunct restaurant, which he had visited earlier. He has no knowledge of the park, or the area, no maps, or hiking experience. He hands over all of his cash to his fiancée, grabs a small bottle of water, and tells her "now I will be able to get ahead of them." and heads directly into the woods, not along any path or trail. It is now 18:00, the fiancée quickly calls 911; her call is routed to the National Park Service and the search begins.

See Instructor's Guide Activity 7-18 and Chapter 10 of *Lost Person Behavior*. This search provides an excellent opportunity to have some discussion of the difficulty of determining the subject's category. Since elements of both despondent and mental illness are present, both must be considered. Options include, choosing only one, giving both equal weight, or giving one scenario more weight than the other. Giving one scenario more weight is easily accomplished by a simple spreadsheet. The user simply enters the distances for two or more subject categories that seem relevant. Then they assign the percentage factor they think is appropriate for each subject category. In this case three different subject categories are considered, despondent, mental illness, and hiker. In reality, hiker is shown simply as an example. The chance the subject is showing behavior similar to a hiker is so low it does not really warrant the 1% chance it is assigned. The bottom line is no correct answer exists for how much weight should be given category. The despondent category cannot be rejected since he did mention the desire to jump off a cliff. Those with mental illness are also associated with increased risk of suicide. Although his plan to drive to a national park with mountains shows some planning, it is weak planning at best. The park lacks significant cliffs that are easily accessible and one would need to use a map to reach them. The lack of medications, worsening

Unit 7 – Subject Categories

hallucinations, and his last comment to his fiancée are more indicative of mental illness. Additional investigative information also indicated that the subject would not be a threat to searchers.

Slide 188

The dashed line shows the path of the car, it then parked straight in (ignoring parking lines) at the lower parking lot. The solid white arrow shows the direction the subject headed into the woods. No path or trails exist at this point.

Slide 189

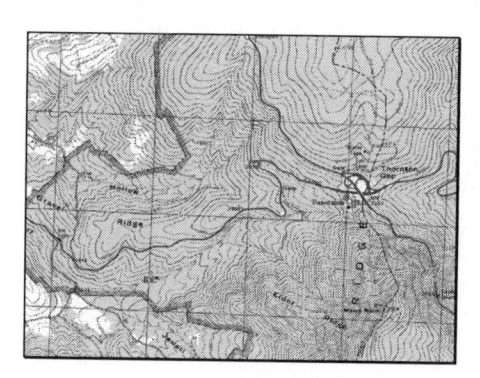

Lost Person Behavior

Unit 7 – Subject Categories

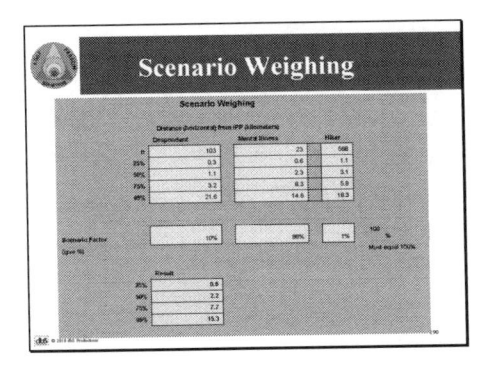

Slide 190

Ideally open the excel spreadsheet to demonstrate how different scenario factors change the results. The excel file is called Scenario Weighing.xls

Slide 191

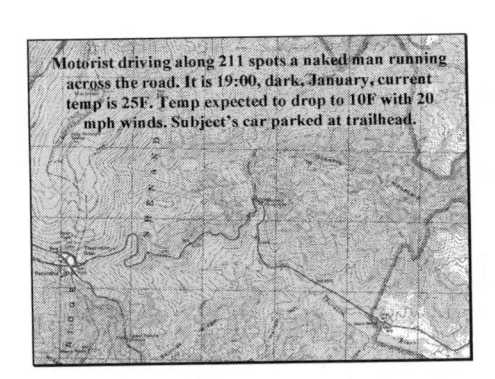

A motorist driving along US 211 which cuts through a national park spots was appears to be at first a deer. Upon closer examination it is a naked man running into the woods. It is 19:00 in January; the current temperature is 25 F. A cold front is moving in and the temperature is expected to drop quickly to 10 F with 20mph winds. The National Park Service is notified along with the State Police. The naked man was spotted at the Turn Bridge Trail trailhead where is vehicle is parked. An examination of his vehicle shows it has been completely torn apart inside. Running the plates reveals it belongs to a 21-year-old male with a history of mental illness. He has no other medical issues. He has not history of depression. He has been off is medicines the last week. His whereabouts are unknown and his parents report he has been acting strangely the last few days. He was last seen heading in the direction of the trail.

Lost Person Behavior   September 2014

Unit 7 – Subject Categories

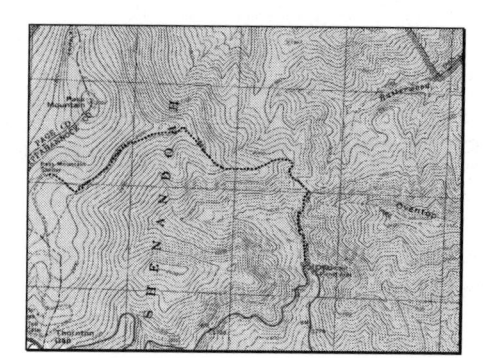

Slide 192

Close up view of the trail. Note that the Oventop fire trail and the Butterwood Fire Trail no longer exist. Only the Pass Mountain trail highlighted by the black line exists. The direction of travel of the subject was North.

Slide 193

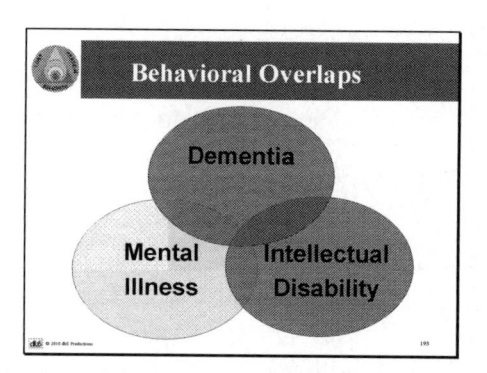

Transition slide between mental retardation and mental illness showing overlap between three different categories.

Slide 194

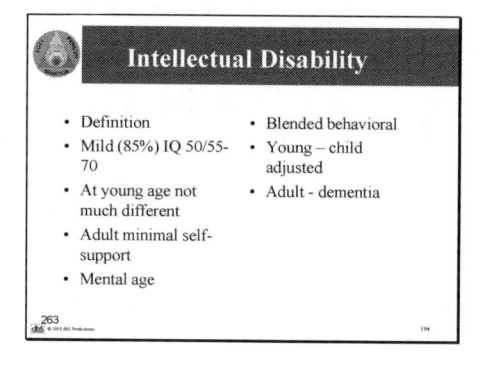

Text reference 206-213 In many of the presentation it may be more appropriate to use the term Intellectual Disability. DSM V will use the term Intellectual disability or ID. The slide heading may be changed if the instructor wishes.

Lost Person Behavior   September 2014

Unit 7 – Subject Categories

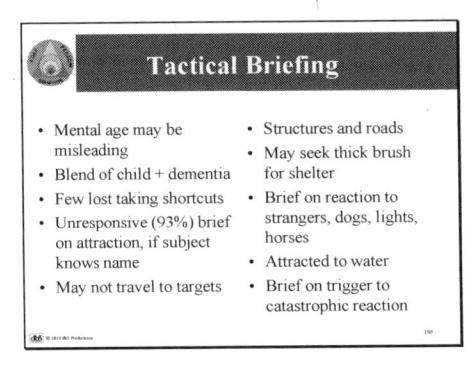

Slide 195

Tactical Briefing

- Mental age may be misleading
- Blend of child + dementia
- Few lost taking shortcuts
- Unresponsive (93%) brief on attraction, if subject knows name
- May not travel to targets
- Structures and roads
- May seek thick brush for shelter
- Brief on reaction to strangers, dogs, lights, horses
- Attracted to water
- Brief on trigger to catastrophic reaction

Slide 196

15-year-old male last seen at 10:00 just outside dorm room. Resident of facility for those with intellectual disability. Temp 45 F, low 25 F. Now 12:00 after search of grounds by Staff.

A 15-year-old male was last seen at 10:00 just outside his dorm room by his counselor after a group outing. They had been playing follow the leader, tag, and ring around the Rosey on the south lawn. They also played on the playground equipment. The counselor had everyone wait at the entrance to the dorm's ramp as he went to open the door. Once everyone had filed in the subject who had been taking up the rear was no where to be found. None of the other children are a reliable witness and no direction of travel can be obtained by a witness. Tracking resources will not be able to obtain a direction of travel either. He is a resident of Lakeland Village which houses those with mental retardation or intellectual disability. He has never runaway or been missing before. He knows his way around the facilities well. He has not been punished or upset over anything. He is alert, attentive, moderately outgoing, and at easy around people, but very shy around strangers. He has no other medical problems. He has normal energy and fitness for a 15-year-old. He is not attracted to anything unusual, but he does like animals.

The current temp is 45 F (7 C) with an expected low of 25 F. It is October and he is dressed in street shoes, lightweight

Unit 7 – Subject Categories

clothing, heavy coat, and a blue stocking cap. It is now 12:00 after a search of the grounds by the staff.

Instructor's Guide Activity 7-20. This map problem is also used in the Managing Land Search Operations course by Skip Stoffel. Several additional resources are available in the Instructor's DVD to the course.

Slide 197

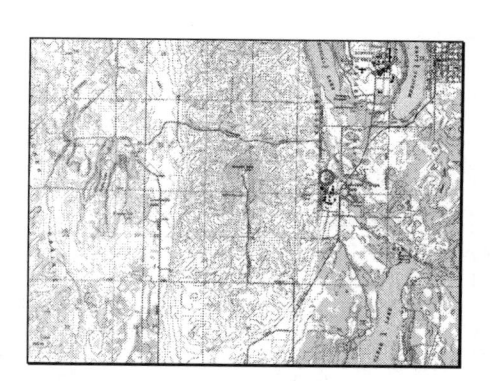

This map should be used for the actual problem. The instructor should point out the IPP, which is just outside his dorm, the facility (Lakeland Village) does not have any gate or fence around it. The top of Booth Hill is fenced and is indicated on the map by the thin red line. The fence has no breaks in it.

Slide 198

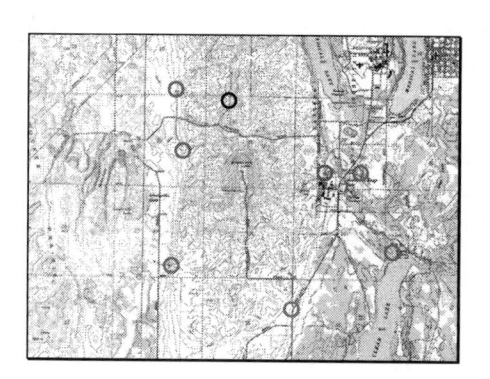

The slide is animated. Each click reveals a location a subject from Lakeland Village has been found. 1) Play area of McDonalds (playing). 2) At the petting zoo – petting animals. 3) At the park playing near the water. 4) Laying in the middle of the road with no clothes. 5) In rancher's barn – burnt it down because he was cold 6,7) In barn – asleep in the hay 16 year old subject of this search. 8) At residence – knocked on door asking for glass of water. 9) Along road, died of hypothermia during the winter.

Unit 7 – Subject Categories

Slide 199

Find Location

	ISRID-Temp	ISRID-Dry	Lakeland-Dry
N	70	10	9
Structure	24%	50%	67%
Road	12%	20%	22%
Linear	2%	10%	
Water	13%		11%
Scrub/brush	9%	10%	
Woods	31%	10%	
Field	7%		

Ask the participants to look their LPB textbook page 209 for the find location chart. Lakeland village is in the dry domain. The cases from Lakeland village were not included in the dry domain ISRID statistics. Both datasets are small and cannot be fully trusted. However, validating one dataset (ISRID) against another (Lakeland) gives some sense of how valid the statistics appear to be. It is unknown how many of the Lakeland village causes may have been severe Autism Spectrum Disorders and how many were pure mental retardation. Among Autism cases we also see 50% found in structures. In the dry domain quick finds prior to SAR activation are often possible in the open areas. Therefore, it is not surprising that more SAR finds are made in structures in dry areas than temperate areas.

Slide 200

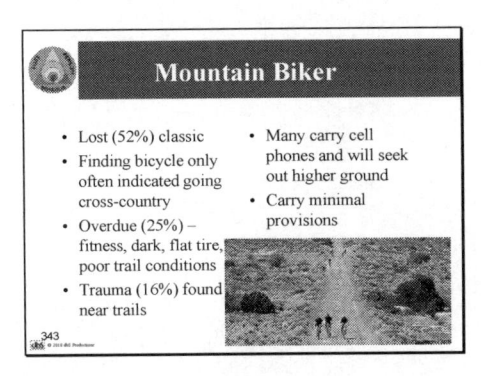

Mountain Biker

- Lost (52%) classic
- Finding bicycle only often indicated going cross-country
- Overdue (25%) – fitness, dark, flat tire, poor trail conditions
- Trauma (16%) found near trails
- Many carry cell phones and will seek out higher ground
- Carry minimal provisions

Text reference: 214-219

Unit 7 – Subject Categories

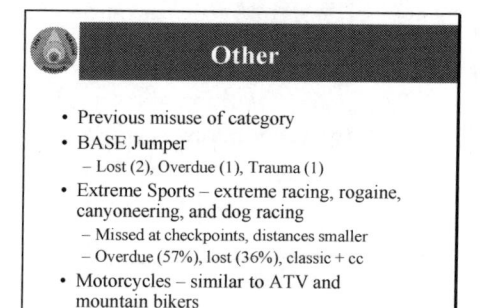

Slide 201

Other
- Previous misuse of category
- BASE Jumper
  - Lost (2), Overdue (1), Trauma (1)
- Extreme Sports – extreme racing, rogaine, canyoneering, and dog racing
  - Missed at checkpoints, distances smaller
  - Overdue (57%), lost (36%), classic + cc
- Motorcycles – similar to ATV and mountain bikers

Text Reference: 220-225. It is important to note that when Syrotuck used the Other category is simply combined all other categories that did not fit. It consisted mostly of gatherers and photographers. However, search planners used this category for anything that did not fit a category. This violates the central premises that the behavior trends of the group will help predict the behavior of the missing individual; if the individual is not at all similar to the group. Other in Lost Person Behavior, is really three distinct groups that did not have sufficient information to warrant an entire sub-chapter. They are BASE jumpers, Extreme Sports, and Motorcycle. In addition, the book provides some additional information on missing parachutes, if the chute opens or not.

Slide 202

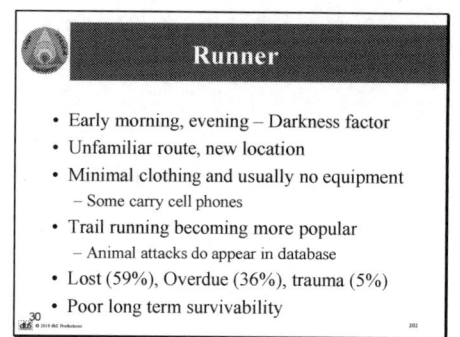

Runner
- Early morning, evening – Darkness factor
- Unfamiliar route, new location
- Minimal clothing and usually no equipment
  - Some carry cell phones
- Trail running becoming more popular
  - Animal attacks do appear in database
- Lost (59%), Overdue (36%), trauma (5%)
- Poor long term survivability

Text Reference: 226-230.

Unit 7 – Subject Categories

Slide 203

Skier - Alpine

- Front country vs. Backcountry
- Well dressed, but minimal equipment.
- Lost (53%): wrong trails, missing trails, going out of bounds common
- Wx often a contributing factor (whiteout)
- Skiers keep moving, build snow shelters, or break into shelters. Moon ↑ % moving
- Trauma (17%)

Text Reference: 231-235. At this point, early indications suggest differences between dry and temperate may be small or not significant. More work will be needed to conduct this analysis. In both cases, lots of snow exists and the vegetation is off the trees in many cases.

Slide 204

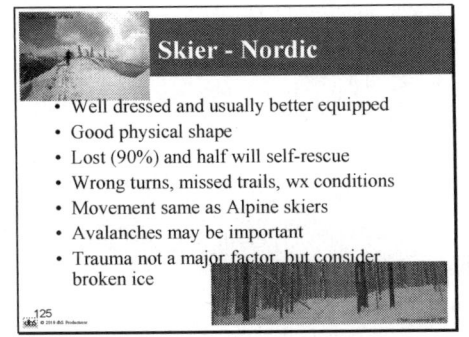

Skier - Nordic

- Well dressed and usually better equipped
- Good physical shape
- Lost (90%) and half will self-rescue
- Wrong turns, missed trails, wx conditions
- Movement same as Alpine skiers
- Avalanches may be important
- Trauma not a major factor, but consider broken ice

Text Reference: 236-241 Note, no difference between dry and temperate seen

Slide 205

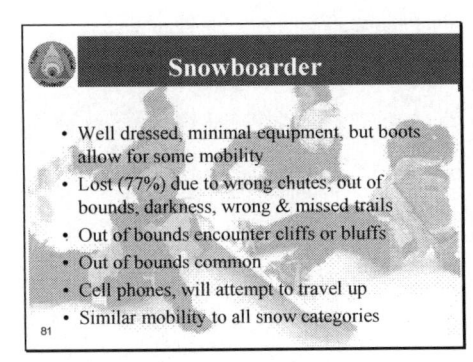

Snowboarder

- Well dressed, minimal equipment, but boots allow for some mobility
- Lost (77%) due to wrong chutes, out of bounds, darkness, wrong & missed trails
- Out of bounds encounter cliffs or bluffs
- Out of bounds common
- Cell phones, will attempt to travel up
- Similar mobility to all snow categories

Text reference: 242-245

Lost Person Behavior — September 2014

Unit 7 – Subject Categories

Slide 206

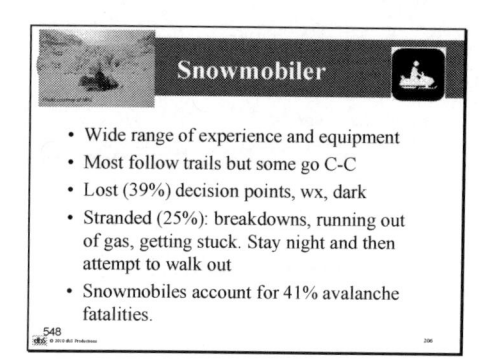

Snowmobiler
- Wide range of experience and equipment
- Most follow trails but some go C-C
- Lost (39%) decision points, wx, dark
- Stranded (25%): breakdowns, running out of gas, getting stuck. Stay night and then attempt to walk out
- Snowmobiles account for 41% avalanche fatalities.

Text reference: 246-251

Slide 207

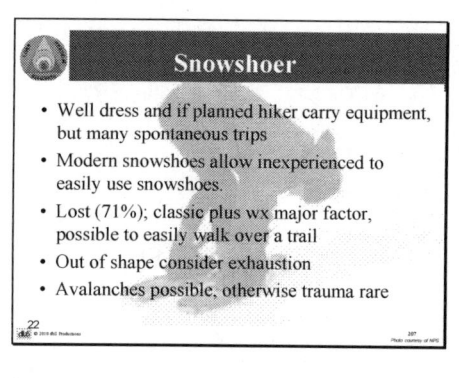

Snowshoer
- Well dress and if planned hiker carry equipment, but many spontaneous trips
- Modern snowshoes allow inexperienced to easily use snowshoes.
- Lost (71%); classic plus wx major factor, possible to easily walk over a trail
- Out of shape consider exhaustion
- Avalanches possible, otherwise trauma rare

Text reference: 252-255

Slide 208

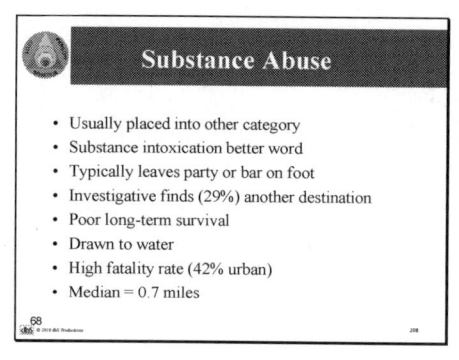

Substance Abuse
- Usually placed into other category
- Substance intoxication better word
- Typically leaves party or bar on foot
- Investigative finds (29%) another destination
- Poor long-term survival
- Drawn to water
- High fatality rate (42% urban)
- Median = 0.7 miles

Text reference: 256-260

Lost Person Behavior — 202 — September 2014

Unit 7 – Subject Categories

Slide 209

You Pick the Category

- A 24-year old male goes camping with his family. Wanders away from camp morning of June 5th. Family reports he was consuming many different types of drugs and drinking large quantities of alcohol.
- A 44-year old male went to Dutchman Peak to look for a lost dog. He was on an ATV with a rifle and alcohol.
- 28-year old male left scene of motor vehicle accident. Alcohol involved.

Slide 210

Subject type? One-year old child or substance abuse? Subject would be classified as child. If actually drank alcohol that would need to be factored into the overall search analysis.

Slide 211

SARA SAR Study

- 100 SAR Fatalities from SARA, AZ (80-92)
- Looked at SAR and medical records
- 50 cases positive findings for alcohol (50%)
- 12 cases positive findings for drugs (12%)
- Alcohol "very probable" causative factor in 40% of unintentional trauma deaths and in 8% of medical related deaths.

The 100 cases were taken from 1980-1992. All were Search & Rescue incidents that Southern Arizona Rescue Association (SARA) responded to in Pima County Arizona.

Lost Person Behavior

Slide 212

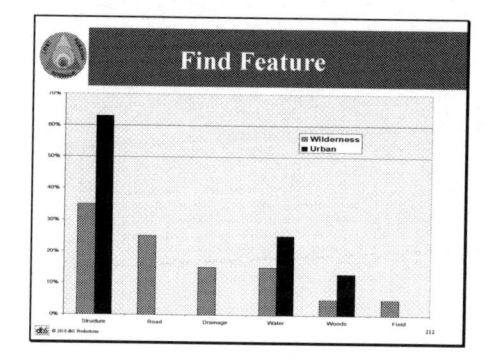

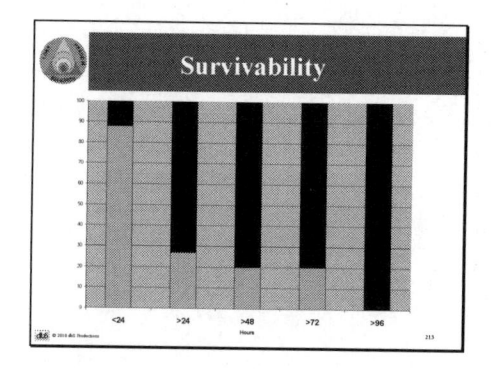

Find Feature

Slide 213

Survivability

Slide 214

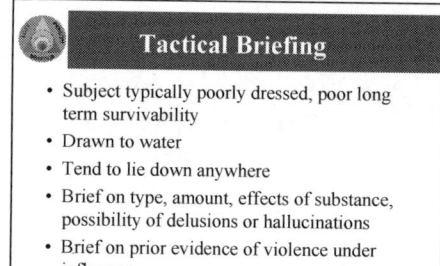

Tactical Briefing
- Subject typically poorly dressed, poor long term survivability
- Drawn to water
- Tend to lie down anywhere
- Brief on type, amount, effects of substance, possibility of delusions or hallucinations
- Brief on prior evidence of violence under influence
- Brief on reaction to authority, police, dogs

Unit 7 – Subject Categories

Slide 215

Incident took place in New Zealand. See Instructor's Activity Guide for additional information. Town of Lyttelton in foreground. Area of bars is indicated by three glasses. Christchurch can be seen at the top of the picture just over the small hills.

Slide 216

Red circle is the bar that marks the IPP. The yellow circle is the location of the friends house where he intended to stay the night.

Slide 217

Red circle is the 25% ring. The yellow circle is the 50% ring. Recommend using this map for students to give input.

Lost Person Behavior

Unit 7 – Subject Categories

Slide 218

Map used to show find location. Subject fell off the road onto the first terrace. And then stumbled off the second terrace resulting in a fatal head wound. Find ring appears on click.

Slide 219

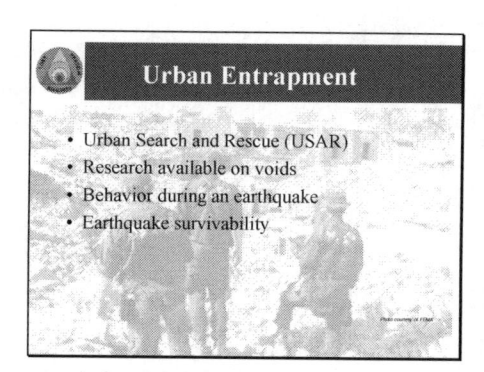

This section is often skipped depending upon the make up of the class. If this area is relevant to participants it may be greatly expanded using information from the book.

Slide 220

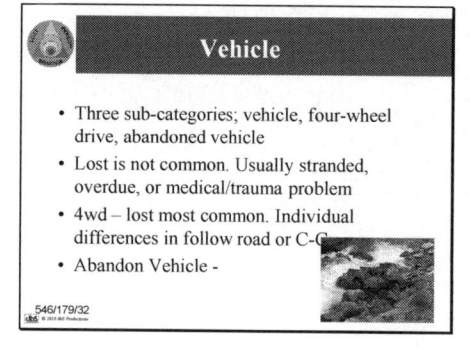

Text Reference: 269-274 Important to note that is category really contains three separate sets of data. In the abandon vehicle sub-cateogry common to see alcohol as a significant contributing factor.

Lost Person Behavior — 206 — September 2014

Unit 7 – Subject Categories

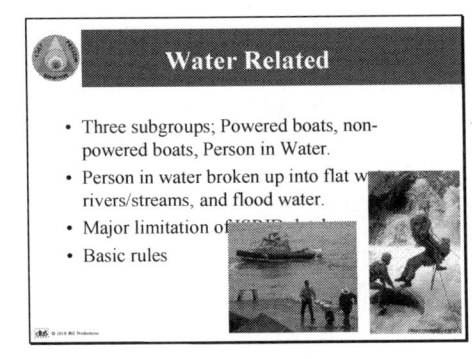

Slide 221

Water Related
- Three subgroups; Powered boats, non-powered boats, Person in Water.
- Person in water broken up into flat w rivers/streams, and flood water.
- Major limitation of ISRID
- Basic rules

Text reference: 275-288. The lecture currently does not provide much information compared to the information available in the text. For the distance from the IPP tables found on page 282 it is important to note that much of the data comes from research compiled by air-scent dog teams trained and used for water search. These searches would represent worst-case scenarios where the initial search failed. In many of these cases the location of the IPP was not correctly determined. Since water related cases make up a large percentage of many SAR teams activates, this section may be enhanced. It should be noted that ISRID was not really designed for water-related research.

Slide 222

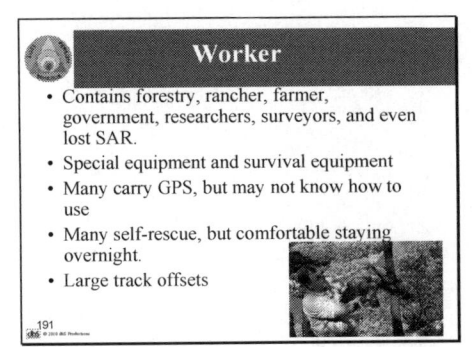

Worker
- Contains forestry, rancher, farmer, government, researchers, surveyors, and even lost SAR.
- Special equipment and survival equipment
- Many carry GPS, but may not know how to use
- Many self-rescue, but comfortable staying overnight.
- Large track offsets

Text reference: 289-293

Lost Person Behavior — 207 — September 2014

Unit 7 – Subject Categories

Slide 223

Support: Better batteries, better clothes, better communications, better map (electronic – remote aerial, update maps – download to GPS/PDA/Phone/Internet device)
The searcher:
Search Management: POA models, computers
The System: Taking the search out of SAR. However, even when given coordinates, still need to search – it is just easier. Technology can take the behavior out of search and rescue but really not the search. Technology can also fail as long as humans and batteries are involved.

Slide 224

Acknowledgements for the creation of the ISRID database.

Unit 7 – Subject Categories

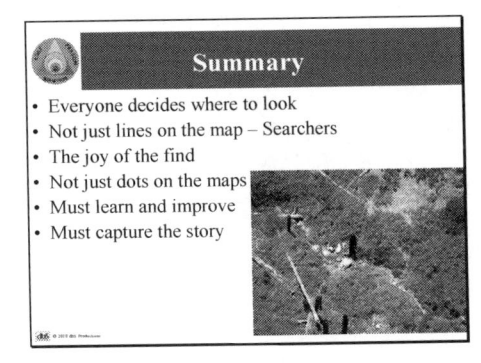

Slide 225

Summary
- Everyone decides where to look
- Not just lines on the map – Searchers
- The joy of the find
- Not just dots on the maps
- Must learn and improve
- Must capture the story

Additional information on the picture can be found at http://www.friendsofyosar.org/rescues/2008/6-8-08_Garmendia_Search.html. The picture was taken by the NPS and shows a subject when he was first found. The search team made extensive use of GIS and profiles.

Slide 226

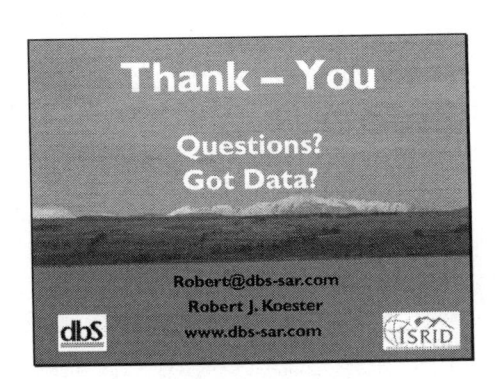

Thank – You
Questions?
Got Data?
Robert@dbs-sar.com
Robert J. Koester
www.dbs-sar.com

Instructor may add their own contact data as well. However, keep contact data for Robert Koester in case students wish to contact.

Unit 7 – Subject Categories

dbS Productions LLC
the source for search and rescue research, publications and training

Instructor Order Form

Online www.dbs-sar.com   **Phone/Fax** +1.434.293.5502   **Mail** PO Box 94 Charlottesville VA 22902

Product	Price (USD)	Quantity	Total
Lost Person Behavior Koester (Mininum order of 12)	$15		
Lost Person Behavior case 32 books	$480		
Lost Person Behavior: Student Workbook	$15		
Lost Person Behavior: Student Workbook license	$3		
Lost Person Behavior Challenge Coin ($10, $8 orders of 12+)	$10/8		
Lost Person Behavior Patches (<12/12-49/>50)	$5/$4/$3		
Urban Search: Young & Wehbring (<12, 12 or >)	$20/$15		
Analysis of Lost Person Behavior Syrotuck	$12		
Vol I. Land Search and Rescue Addendum	$18		
Vol II. Catastrophic Incident Search and Rescue Addendum	$18		
Vol III Incident Command System Field Operations Guide for SAR	$20		
Lost Alzheimer's Disease Search Management Koester	$40		
Man-Trackers and Dog Handlers in SAR Fuller, Johnson, Koester	$10		
Fatigue: Sleep Management During Disasters and Sustained Operations	$10		
Incident Commander for Ground Search and Rescue - Instructor	$75		
SAR Skills for the Emergency Responder Stoffel	$41		
The Textbook for Managing Land Search Operations Stoffel	$36		
The Handbook for Managing Land Search Operations Stoffel	$28		
The Handbook for Aviation Survival Sense Stoffel	$35		
The 98.6 Survival Kit ERI	$90		
Search Wheel (Temperate, Dry, Urban) (Coming Soon)	$15		

	SUBTOTAL	
	Tax (5% for Virginia residents only)	
	actual cost to ship + handling	
	TOTAL	

Name	Organization	
Billing Address		Address 2
City	State	Zip
Shipping Address		Address 2
City	State	Zip
Province/Region	Country	Postal Code
E-mail		Phone
Card #		
Card Type	Exp Date	Verification #

Questions on orders? Please call 800.745.1581 or +1.434.293.5502 or e-mail emily@dbs-sar.com.

Made in the USA
Columbia, SC
28 September 2024